T0181991

Geometry - Intuition and Concepts

Jost-Hinrich Eschenburg

Geometry - Intuition and Concepts

Imagining, understanding, thinking beyond. An introduction for students

Jost-Hinrich Eschenburg
Institut für Mathematik
Universität Augsburg
Augsburg, Germany

ISBN 978-3-658-38639-9 ISBN 978-3-658-38640-5 (eBook)
https://doi.org/10.1007/978-3-658-38640-5
The translation was done with the help of artificial intelligence (machine translation by the service DeepL.com). A subsequent human revision was done primarily in terms of content.

This Springer imprint is published by the registered company Springer Fachmedien Wiesbaden GmbH, part of Springer Nature.
The registered company address is: Abraham-Lincoln-Str. 46, 65189 Wiesbaden, Germany

Contents

1 **What Is Geometry?** 1

2 **Parallelism: Affine Geometry** 7
- 2.1 From Affine Geometry to Linear Algebra 7
- 2.2 Definition of the Affine Space 11
- 2.3 Parallel and Semi-Affine Mappings 13
- 2.4 Parallel Projections 16
- 2.5 Affine Representations, Ratio, Center of Gravity 18

3 **Incidence: Projective Geometry** 21
- 3.1 Central Perspective 21
- 3.2 Points at Infinity and Projection Lines 25
- 3.3 Projective and Affine Space 26
- 3.4 Semiprojective Mappings and Collineations 29
- 3.5 Theorem of Desargues 33
- 3.6 Conic Sections and Quadrics; Homogenization 36
- 3.7 Theorem of Brianchon 42
- 3.8 Duality and Polarity; Pascal's Theorem 45
- 3.9 Projective Determination of Quadrics 49
- 3.10 The Cross-Ratio 52

4 **Distance: Euclidean Geometry** 57
- 4.1 The Pythagorean Theorem 57
- 4.2 The Scalar Product in $\mathbb{R}^n$ 61
- 4.3 Isometries of Euclidean Space 65
- 4.4 Classification of Isometries 66
- 4.5 Platonic Solids 69
- 4.6 Symmetry Groups of Platonic Solids 73
- 4.7 Finite Subgroups of the Orthogonal Group, Patterns, and Crystals 76
- 4.8 Metric Properties of Conic Sections 79

5 **Curvature: Differential Geometry** 85
- 5.1 Smoothness 85
- 5.2 Fundamental Forms and Curvatures 87

5.3 Characterization of Spheres and Hyperplanes 90
5.4 Orthogonal Hypersurface Systems 91

6 Angle: Conformal Geometry 95
6.1 Conformal Mappings 95
6.2 Inversions 97
6.3 Conformal and Spherical Mappings 99
6.4 The Stereographic Projection 100
6.5 The Space of Spheres 103
6.6 Möbius and Lie Geometry of Spheres 104

7 Angular Distance: Spherical and Hyperbolic Geometry 107
7.1 Hyperbolic Space 107
7.2 Distance on the Sphere and in Hyperbolic Space 109
7.3 Models of Hyperbolic Geometry 112

8 Exercises 115
8.1 Affine Geometry (Chap. 2) 115
8.2 Projective Geometry (Chap. 3) 120
8.3 Euclidean Geometry (Chap. 4) 125
8.4 Differential Geometry (Chap. 5) 132
8.5 Conformal Geometry (Chap. 6) 133
8.6 Spherical and Hyperbolic Geometry (Chap. 7) 136

9 Solutions 139

Literature (Small Selection) 161

Index 163

1 What Is Geometry?

Abstract

This introductory chapter has rather a philosophical, or more precisely a metamathematical character: It does not do mathematics, but talks about mathematics. We try to work out what "geometry" actually is, given the many "geometries" that will be talked about in our book, how we gain geometric knowledge and what its scope of application is. It is also about the second part of our title, Intuition and Concepts. Where do they each have their place and what is their relation to each other? In particular, we will talk about the importance of axiomatics, which has been the starting point of all mathematical reasoning since the beginning of the twentieth century, and we will give reasons why we no longer see any need for a separate axiomatic foundation for geometry today.

The word "geometry" comes from the Greek and actually means earth-survey. Geometric knowledge existed in all cultures, but it was only in ancient Greece that it became a science in the sense we know it today: a systematic way of gaining secure knowledge. In its original sense, the content of geometry is the study of spatial forms.[1] However, the ideas and linguistic means developed in this context can be transferred beyond applications in visible space to other problem areas; an example already familiar from the beginners' lectures is n-dimensional space. Geometry in the present sense of the word is the consideration of mathematics from the point of view of this circle of ideas derived from the spatial perspective.

Let us start by explaining the words of the title, intuition and concept. What does "Intuition" mean and what role does it play in mathematics, especially in geometry? Intuition in geometry concerns first of all the spatial forms (shapes) present in *reality*

[1] Thus it has a strong relation to the fine arts, which we also do not wish to neglect altogether, see, e.g., Sects. 3.1 and 4.7.

J.-H. Eschenburg, *Geometry – Intuition and Concepts*,
https://doi.org/10.1007/978-3-658-38640-5_1

and their apparent or hidden relations. The forms, however, are not simply taken from reality, but they are *idealized,* transformed into an *idea* in the sense of Plato. In reality there are more or less circular objects; in our imagination the idea of the (perfect) circle is formed from them, and this already happens with preschool children. Finally, *mathematics* puts the idea into words; the circle becomes the *set of all points of the plane that have a constant distance ("radius") from a fixed point ("center").* The relation between idea and mathematical formalization as a *definition* should be perfect, i.e. it should put exactly this idea into words, no more and no less:

$$\text{Reality} \longrightarrow \text{Idea} \longleftrightarrow \text{Mathematical Concept} \tag{1.1}$$

The idea is thus embedded in a certain conceptual framework and made accessible to further processing by logic. The example of the circle, however, also shows the problematic nature of this procedure. The idea of the circle, familiar to every child, becomes an unwieldy formulation containing new words which themselves have to be explained again: set, points, plane, distance. Moreover, the definition by no means expresses everything that is included in the word "circle", e.g. the uniform roundness *(curvature).* Further terms *(curve, second derivative)* are needed to translate this aspect into mathematical language. On the other hand, without the conceptual penetration, we cannot arrive at secure knowledge, because the view (the "appearance") can be deceptive.[2]

This has both educational and mathematical consequences. Should one, for example, sacrifice the intuitive grasp of an idea in all its aspects ("the circle is round") in order to arrive at the same result after a long analysis? This would certainly not be a gain, because intuitive recognition is of great value. But there are also situations for the understanding of which intuition is no longer sufficient and which require a precise definition, for example when we want to determine the points of intersection of two circles; there the value of formalization becomes apparent. We mathematicians, on the other hand, are just like schoolchildren when we are faced with unsolved problems. In such a situation, the means of finding in geometry is often not the logical conclusion, but the *figure,* which is actually (as a drawing on the paper, the blackboard or the computer screen) located on the left ("reality") of our scheme (1.1), but which is supposed to symbolize an ideal situation. The realization of the *hidden* through its traceability to the *obvious* (this tracing back is called *proof*) happens in its essential part by means of the figure, in the simplest case by introducing suitable auxiliary lines. As an example of the "force

[2] "It comes along with our nature that *intuition* can never be other than *sensible*, i.e., that it contains only the way in which we are affected by objects. The faculty for *thinking* of objects of sensible intuition, on the contrary, is the *understanding*. Neither of these properties is to be preferred to the other. Without sensibility no object would be given to us, and without understanding none would be thought. Thoughts without content are empty, intuitions without concepts are blind." (Immanuel Kant, 1724–1804, Critique of Pure Reason, A51). https://cpb-us-w2.wpmucdn.com/u.osu.edu/dist/5/25851/files/2017/09/kant-first-critique-cambridge-1m89prv.pdf.

of the figure" let us consider the determination of the sum of angles in a triangle by introducing the parallels,

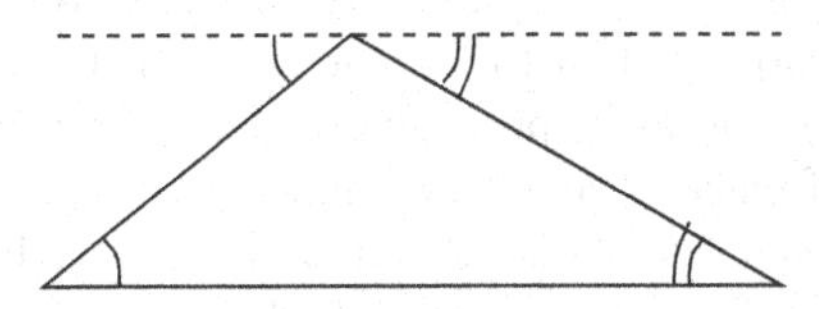

or the proof of Pythagoras' theorem[3] by introducing the oblique square:

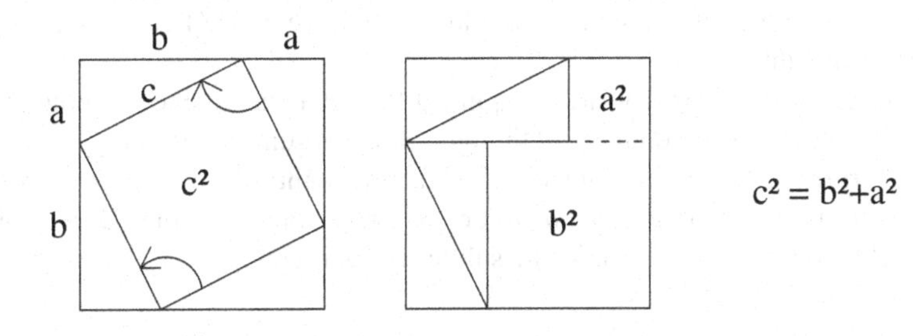

The transfer of the proof to the right-hand side in our scheme (1.1), the *formalization,* is thereafter an automatic process, often rightly omitted because it is too tedious and of little value.

But what is to count as the "obvious"? We can agree on this in each case. In the figure for the sum of the angles of the triangle, for example, the equality of the alternate angles at parallels should be "obvious", because a rotation of the figure makes one merge into the other. This is actually a *symmetry consideration*: The angles are equal because there is an angle-preserving mapping (a *symmetry mapping*) connecting them. Such arguments have strong intuitive force because we know rotations, displacements, and reflections very well from everyday experience. Incidentally, in the Pythagorean figure with the oblique square, it is not entirely obvious why the area of a figure remains unchanged with rotations; we will discuss this in Sect. 4.1.

Mathematicians have undertaken this unification process once and for all by agreeing on *axioms*, mathematical statements on which they based all further conclusions. This was first done in Euclid's "Elements", which summarized the geometrical knowledge of the time around 300 BC.[4] In a more modern form this task was accomplished for geometry in the "Foundations of Geometry" [9] by David Hilbert (1899).[5] This book triggered a whole wave of *axiomatizations*

[3] Pythagoras of Samos, c. 569–500 BC.

[4] Euclid of Alexandria, ca. 325–265 BC.

[5] David Hilbert, 1862 (Königsberg, now Kaliningrad)–1943 (Göttingen).

that eventually covered all areas of mathematics; for calculus (axioms of the real numbers) and linear algebra (vector space axioms) you learned this in the basic lectures.[6] We want to rely on this knowledge now as well. So we do not want to set up a separate axiom system for geometry, even if we occasionally discuss the (extremely simple) axioms of projective geometry, but prefer to fall back on the familiar axioms of calculus and linear algebra. The mathematical conceptual framework, the right-hand side of our scheme, is already ready-made; we only need to relate geometry to it. To this end, we will show how geometric concepts and facts can be translated into the language of these areas. On the one hand, we would like to strengthen the bridge from university mathematics to everyday and school geometry, and on the other hand, we would like to have a safe framework for new geometrical insights.

The concepts of geometry are of quite different natures; they denote, so to speak, different layers of geometrical thought: Some arguments use only terms such as *point, straight line*, and *incidence* (the statement that a given point lies on a given straight line), others use *distance*- or *symmetry*-considerations. Each of these conceptual fields determines a separate subfield of geometry:

- Incidence: Projective geometry
- Parallelism: Affine geometry
- Angle: Conformal geometry
- Distance: Metric geometry
- Curvature: Differential geometry
- Angular distance: Spherical and Hyperbolic geometry
- Symmetry: Mapping geometry

These geometric areas are intertwined: Straight line segments minimize the distance between their endpoints and have zero curvature, parallels have constant distance, distances also determine angles, etc. The last conceptual environment "symmetry" runs through all the others; *Felix Klein* in his "Erlangen program" of 1872 drew attention to the relations between geometry and symmetry groups.[7] This list of subfields of geometry is not complete if dimensions higher than three are allowed;

[6] In the modern sense, axioms are not, as in Euclid's work, fundamental true statements that are immediately obvious and therefore require no proof, but rather they are the defining properties of a concept or a domain of mathematics: A vector space, for example, is a structure that satisfies the vector space axioms.

[7] Felix Klein, 1849 (Düsseldorf)–1925 (Göttingen), https://arxiv.org/pdf/0807.3161.pdf.

in some sense it was *J. Tits*[8] who found the complete list around 1960[9] (after preliminary work by *W. Killing, S. Lie, E. Cartan, H. Weyl*[10] and others).

At each stage we will start with the left-hand side of our scheme (1.1), with visualization, and we will use whatever we know (from wherever) of visual geometry. This will lead us to an embedding of the facts into our mathematical model. Only in this framework we will give mathematically exact definitions and proofs. We will start with affine geometry, since it is more familiar to you than projective geometry.

As most important literature I will mention two books: First the beautiful book by D. Hilbert and S. Cohn-Vossen: "Geometry and the Imagination" [2], which was first published in 1932; a mathematics book almost without formulas, but with all the more pictures. Much more comprehensive is "Geometry" by Marcel Berger [1]; characteristic for this book is that the geometrical arguments are often only indicated; you have to think about them to carry them out, but then you will see that all the essential information for them has been given.

The book grew out of a one-semester series of lectures at the University of Augsburg, which was revised several times. It is my pleasure to give special thanks to Dr. Erich Dorner, who read the manuscript over and over again with great patience and pointed out countless errors to me.

[8] Jaques Tits, 1930 (Uccle/Ukkel, Belgium)–2021 (Paris).

[9] cf. J. Tits: *Buildings of spherical type and finite BN-pairs,* Springer Lecture Notes in Math. 386 (1974).

[10] Wilhelm Killing, 1847 (Burbach)–1923 (Münster), Sophus Lie, 1842 (Nordfjordeid)–1899 (Oslo), Élie Cartan, 1869 (Dolomieu)–1951 (Paris), Hermann Weyl, 1885 (Elmshorn)–1955 (Zürich).

2 Parallelism: Affine Geometry

Abstract

Straight line and incidence are the simplest geometric notions. In our opening chapter, however, we will add the notion of parallelism, which will later be recognized as a special case of incidence. This brings us to affine geometry, which is very close to our vision. Even more important: From it, linear algebra can be founded descriptively, because descriptive vector addition has to do with parallelograms, scalar multiplication with homotheties and ray theorems. Thus, in the second step, we can embed affine geometry into linear algebra and express geometric facts algebraically. This concerns especially the symmetry group, the group of all transformations that preserve straight lines and parallels: We can characterize them algebraically. The algebraic point of view allows us, without additional effort, to go beyond our spatial intuition in two respects, and thus to apply geometry to non-visual situations: The number of dimensions may be arbitrary, even larger than two or three, and the field of real numbers describing the one-dimensional continuum may be replaced by an arbitrary field.

2.1 From Affine Geometry to Linear Algebra

The *affine geometry* is distilling precisely the vector space structure from the well known descriptive geometry of the plane or space (both of which we denote by X). Its basic notions are point, straight line, incidence and parallelism. As announced in the introduction, in this subsection we will not yet do mathematics in today's understanding, the starting point of which are the axioms; we first want to develop these from geometric vision. Only from the next subsection on will we work within the fixed conceptual framework of linear algebra.

The basic idea for the development of linear algebra from visual observation is the *parallelogram construction*: We fix a point $o \in X$ arbitrarily and call it *origin*.

J.-H. Eschenburg, *Geometry – Intuition and Concepts*,
https://doi.org/10.1007/978-3-658-38640-5_2

If now two other points $x, y \in X$ are given such that o, x, y do not lie on a common straight line (not *collinear*), then we denote the fourth vertex of the parallelogram with vertices o, x, y as $x + y$.

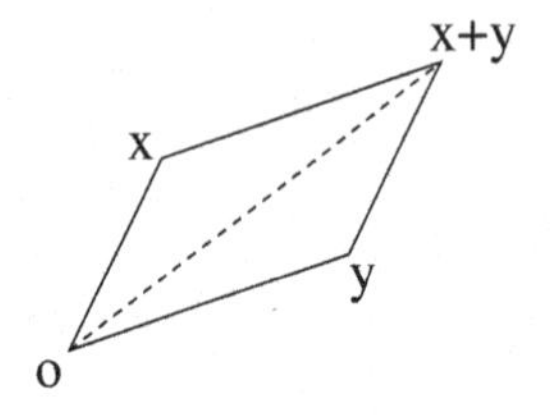

Obviously, this operation is commutative, $x + y = y + x$, and also associative, $(x + y) + z = x + (y + z)$ as the following figure shows:

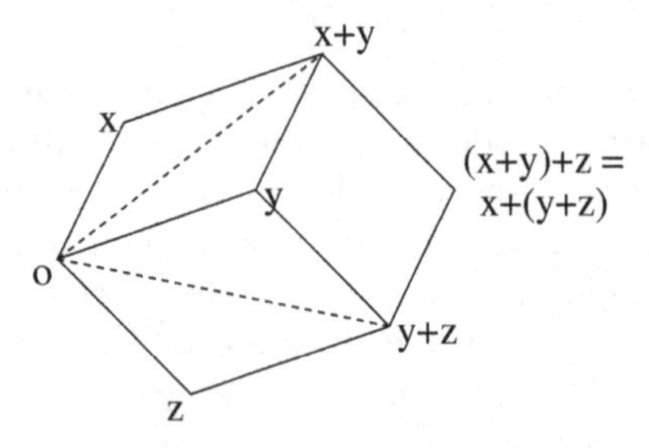

There is another way to describe the construction: The point $x + y$ is the endpoint of the line segment, which is created by parallel displacement of the line ox to the new starting point y (or vice versa by moving oy into the new starting point x). This description has the advantage that it can also be applied to *collinear* (lying on a common straight line) points o, x, y; it corresponds to the geometric addition of values on a scale, that is, to the addition of numbers.

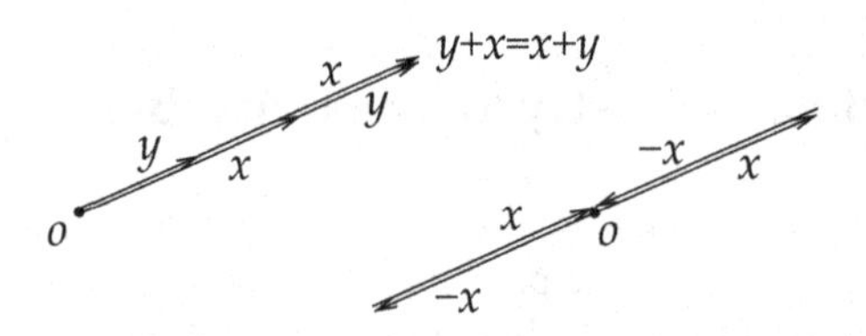

In particular, we find a point located on the straight line ox at the same distance from o as x, but on the other side of o. We call this point $-x$ because $x + (-x) = o$ (right figure). Thus $(X, +)$ becomes an *abelian group,* where the point o plays the role of the neutral element 0, that is: $x + o = x = o + x$. The directed line segment $\vec{ox}$ will be called *vector* and the operation just described the *vector addition.*

In order to make X a *vector space* over the field $\mathbb{R}$ (the real numbers), we additionally have to see the multiplications by *scalars,* that is, with real numbers.

They are called *homotheties* in geometry. From the addition we readily get the multiplications by integers: $2x = x + x$, $3x = x + x + x$, $(-2)x = (-x) + (-x)$. Multiplications by rational numbers (fractions) are defined conversely: $y = \frac{1}{3}x$ is the point on the straight line ox with $3y = x$, analogously $\frac{1}{n}x$ for any $n \in \mathbb{N}$ and $\frac{m}{n}x = m(\frac{1}{n}x)$ for all $m \in \mathbb{Z}$. The multiplication with any $\lambda \in \mathbb{R}$ is done by approximation of λ by fractions.

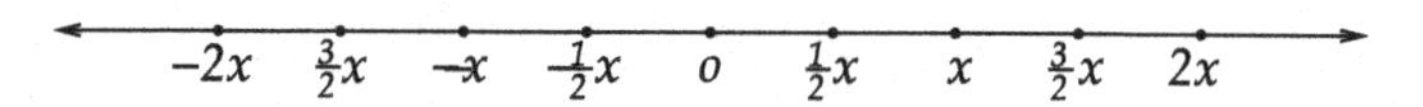

The straight line ox thus becomes an isomorphic image of the number line $\mathbb{R}$. Thus vector space axioms (ii) and (iv) hold.[1] Two homotheties with factors λ and μ combine to a common homothety with the factor $\lambda\mu$, therefore vector space axiom (iii) holds, $\lambda(\mu x) = (\lambda\mu)x$.

Each homothety is already determined by its action on a single point $x \neq o$; its construction is given by the "ray theorem" of school geometry, which describes the phenomenon of enlargement along rays through a common point:

If rays are intersected by parallels, the segments on the rays behave like the segments on the parallels.

In particular, the scaling factor on all rays is the same, and we thus obtain a geometric construction of homothety:

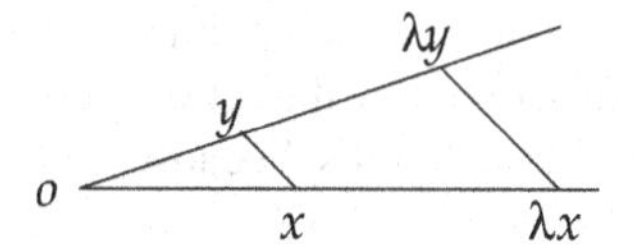

[1] As a reminder: A *field* is a set $\mathbb{K}$ with two distinct distinguished elements $0, 1 \in \mathbb{K}$ and two mappings $+, \cdot : \mathbb{K} \times \mathbb{K} \to \mathbb{K}$ with the following properties (*field axioms*):

(i) $(\mathbb{K}, 0, +)$ is a commutative group,
(ii) $(\mathbb{K}^*, 1, \cdot)$ is a commutative group, where $\mathbb{K}^* = \mathbb{K} \setminus \{0\}$,
(iii) $\alpha(\beta + \gamma) = \alpha\beta + \alpha\gamma$ for all $\alpha, \beta, \gamma \in \mathbb{K}$.

Sometimes one omits the commutativity of the multiplicative group $\mathbb{K}^*$ in which case $\mathbb{K}$ is called a *skew field*. Further, a *vector space* over a (skew) field $\mathbb{K}$ is an abelian group $(V, +)$ with a mapping $\mathbb{K} \times V \to V$, $(\lambda, v) \mapsto \lambda \cdot v = \lambda v$, called *scalar multiplication*, with the following properties (*vector space axioms*):

(i) $\lambda(v + w) = \lambda v + \lambda w$,
(ii) $(\lambda + \mu)v = \lambda v + \mu v$,
(iii) $(\lambda\mu)v = \lambda(\mu v)$,
(iv) $1 \cdot v = v$, for all $\lambda, \mu \in \mathbb{K}$ and $v, w \in V$.

By means of a basis any finitely generated vector space over $\mathbb{K}$ can be identified with $\mathbb{K}^n = \{(x_1, \ldots, x_n);\ x_1, \ldots, x_n \in \mathbb{K}\}$ for some $n \in \mathbb{N}$.

If the points x and λx on the straight line ox are given, then λy is the intersection of the line oy with the parallel to the straight line xy through the point λx. Thus we also obtain the still missing vector space axiom (i) (see Footnote 1), the distributive law $\lambda(x + y) = \lambda x + \lambda y$ for $\lambda > 0$, with the help of the following figure:

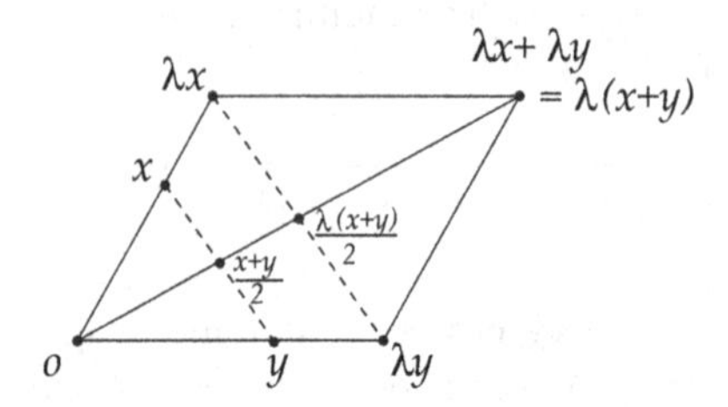

The construction of the vector space structure from descriptive geometry relies on two kinds of mappings on X: the *translations*, or *parallel displacements* on the one hand and the *homotheties* on the other hand. Both are *collineations*, i.e. they transform straight lines into straight lines. Additionally they *preserve directions*,[2] i.e. every straight line transforms into a straight line parallel to it. Such maps are called *dilatations*. In our algebraic formalism these are the mappings

$$T_v : X \to X, \quad x \mapsto v + x, \tag{2.1}$$

$$S_\lambda : X \to X, \quad x \mapsto \lambda x \tag{2.2}$$

for given $v \in X$ and $\lambda \in \mathbb{R}$. From the viewpoint of geometry one can say that the vector space belonging to X is formed by the set of translations T_v. The vector space addition is the composition of translations, because

$$T_v \circ T_w = T_{v+w}, \tag{2.3}$$

as the parallelogram construction shows. The multiplication with the scalar λ corresponds (under the map $v \mapsto T_v$) to the conjugation of the translation T_v with the homothety S_λ,

$$T_{\lambda v} = S_\lambda \circ T_v \circ S_\lambda^{-1}, \tag{2.4}$$

since $S_\lambda(T_v(S_\lambda^{-1} x)) = \lambda(\lambda^{-1} x + v) = x + \lambda v = T_{\lambda v} x$. A composition of homotheties corresponds to the multiplication of scaling factors:

$$S_\lambda \circ S_\mu = S_{\lambda\mu}. \tag{2.5}$$

[2] A *direction* in X is a class of parallel straight lines.

Clearly, compositions of translations and homotheties form a group, the group of *dilatations.*

2.2 Definition of the Affine Space

We have reconstructed linear algebra from affine geometry (the geometry of straight lines and parallels); affine geometry thus takes place in a vector space. However, our labeling of the point o was quite arbitrary; we could just as well have chosen any other point of X as the origin. We therefore define (somewhat imprecisely at first) an *affine space* as "a vector space X without distinction of origin 0" (whatever that means exactly). In doing so, we go beyond our geometric intuition in two aspects:

- The dimension n of X can be arbitrary, not just 2 or 3,
- X can be a vector space over an *arbitrary field* $\mathbb{K}$, not only over $\mathbb{R}$. We think for instance of $\mathbb{K} = \mathbb{C}$ or $\mathbb{K} = \mathbb{C}(z)$ (the field of rational functions in one complex variable z) or $\mathbb{K} = \mathbb{F}_p$ (the field with p elements for a prime number p) or its algebraic extensions $\mathbb{K} = \mathbb{F}_{p^n}$.[3] Sometimes we even choose $\mathbb{K}$ to be a *skew field* for which the multiplication is no longer commutative; cf. Exercise 4. An important example are the *quaternions,* to which we will return on various occasions; cf. Exercise 34.

Here we meet the more general definition of *geometry*: We use the ideas of visual intuition developed in plane and space in order to understand relationships that are no longer intuitive, such as the structure of $\mathbb{K}^n$.

The missing distinction of the zero point (origin) becomes apparent in the concept of *affine subspaces* since these generally do not contain this point at all: A k-dimensional *affine subspace* of X is a subset of the form $U+x = \{u+x;\ u \in U\}$, where $U \subset X$ is a k-dimensional *linear subspace* of X.[4] Moreover, two affine subspaces of the form $U + x$ and $U + y$ for the same linear space U are called *parallel.* Through each point $x \in X$ passes exactly one of the parallel affine subspaces $U + x$, and $U + x$ passes through 0 exactly if $x \in U$, which means $U + x = U$.

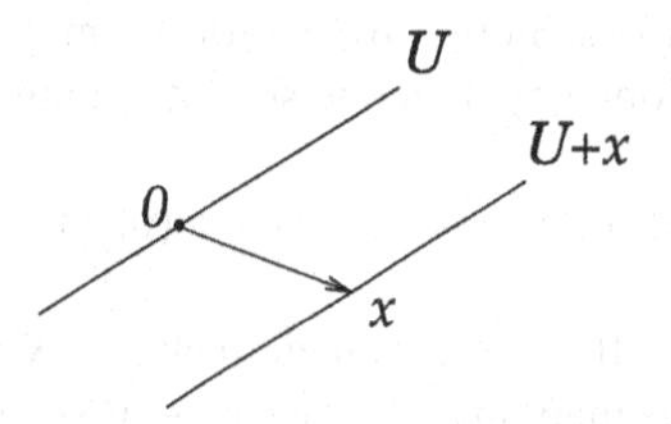

[3] However, we have to make a restriction in several places: We must have $1+1 \neq 0$ in $\mathbb{K}$, otherwise we cannot divide by 2. This is violated in $\mathbb{F}_2$ and all its field extensions. The smallest number p for which the p-fold sum of 1 is zero is called the *characteristic* of the field $\mathbb{K}$. The fields of characteristic 2 more often play a special role and must be excluded in the general argument.

[4] $0 \in U, u + u' \in U, \lambda u \in U$ for all $u, u' \in U$ and $\lambda \in \mathbb{K}$.

Affine subspaces of dimensions $k = 0, 1, 2, n-1$ (for $n = \dim X$) are called *points, straight lines, planes,* and *hyperplanes.* Thus we have expressed the basic concepts of affine geometry (points, lines, etc., and parallelism) by those of linear algebra.

However, the definition of affine space still has a flaw: What should "without distinction of the zero point" mean? And even worse: An affine subspace should be an affine space itself, but it does not contain the zero point, and how can one "not distinguish" a point which is not in it at all? The "correct" definition avoids these difficulties; it reads:

> **Definition:** An *affine space* is a set X on which a *vector group V operates simply transitively.*

As always, we must pay for greater precision by introducing new terminology: A *vector group* is the commutative additive group $(V, +)$ of a vector space V. In geometric terms it is the translation group on V, see (2.3). A group $(V, +)$ *operates* on a set X if there is a mapping $w : V \times X \to X$ (called *operation* or *action* of the group $(V, +)$ on X) with the following two properties.

$$w(0, x) = x, \quad w(a + b, x) = w(a, w(b, x)) \tag{2.6}$$

for all $a, b \in V$ and $x \in X$. In particular $w(a, w(-a, x)) = w(0, x) = x$ and therefore for each $a \in V$ the mapping

$$w_a : X \to X, \ x \mapsto w(a, x)$$

is bijective with inverse mapping w_{-a}. Therefore, we can view the action w also as a mapping $w : V \to B(X)$, $a \mapsto w_a$ into the group $B(X)$ of the bijective mappings on X (with the composition as group operation), and Eq. (2.6) says precisely that w is a *homomorphism of groups*:

$$w_0 = \mathrm{id}_X, \quad w_a w_b = w_{a+b} \tag{2.7}$$

for all $a, b \in V$. A group action w from V to X is called *transitive,* if any two point $x, y \in X$ are mapped onto each other by one of the mappings w_a, and it is called *simply transitive* if this is done by only one such mapping, i.e. if the mapping

$$w^x : V \to X, \ v \mapsto w(v, x)$$

is bijective for each $x \in X$. If we select an element $o \in X$ we can therefore identify X and V using the bijective mapping w^o where $0 \in V$ is mapped to $o \in X$.

A vector space V is also an affine space in this sense, since $(V, +)$ operates on $X = V$ simply transitive by $w(a, x) = a+x$. In this case w_a is the translation T_a; we will therefore prefer to call this particular action T instead of w. It is in fact simply transitive, because any two points x, y can be connected by exactly one vector a, namely $a = y - x$ (thus $y = x + a = T_a x$). Also every affine subspace $U + x \subset V$

is an affine space according to the new definition, because the vector group $(U, +)$ operates on it simply transitively by

$$U \times (U + x) \ni (u, u' + x) \mapsto u + u' + x \in U + x.$$

The subgroup $(U, +) \subset (V, +)$ operates on all of V, namely by the restriction $T|_U$ of the action $T : V \to B(V)$, and the subsets $U + x$ are *invariant* under $T|_U$, i.e. the elements of $x + U$ are mapped again to $x + U$ by the mappings T_u, $u \in U$. The action $T|_U$ on V is no longer transitive, therefore V decomposes into a disjoint union of *transitivity domains* or *orbits* under the action of U: These are the parallel affine subspaces $U + x$, $x \in V$ (cf. Exercise 2).

We will always assume in the following that our affine space X is a vector space, i.e., that an origin $o \in X$ has been chosen. However, we will make clear in any statement of affine geometry that it is independent of the choice of o, so it is preserved when we apply a translation. Translations are again just a special case of parallel mappings, which we will study in the following section.

2.3 Parallel and Semi-Affine Mappings

We further consider a vector space X, which we take to be an affine space; the field $\mathbb{K}$ may be arbitrary. Straight lines and parallels are the fundamental notions of affine geometry; therefore the automorphisms or symmetries of affine geometry are precisely the invertible mappings $F : X \to X$ which preserve straight lines and parallels, that is, they map straight lines bijectively to straight lines and parallels to parallels. We will call them *parallel maps*.[5] We want to transform this geometric description into an algebraic one. To do so we will first consider only those parallel mappings F, which additionally fix the origin: $F(o) = o$. A parallelogram spanned by any two vectors x, y is transformed by F into the parallelogram spanned by $F(x)$ and $F(y)$. Thus $F(x) + F(y)$ is the F-image of $x + y$ and therefore F is *additive:* $F(x + y) = F(x) + F(y)$ for all $x, y \in X$.

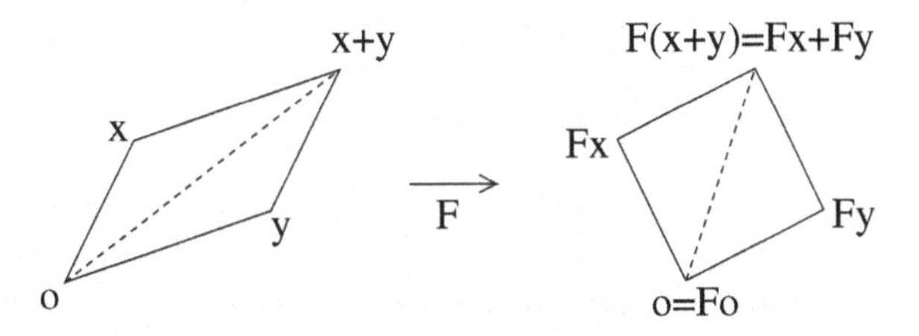

[5] We must assume $\dim X \geq 2$ in the following. In dimension 1, there are no parallels, and the term "parallel map" makes no sense. However, even in dimension 1 we still have the notion of parallel shift and hence line segment addition; this must be respected by an affine mapping F which replaces the notion of parallelity, see Sect. 2.5. We see here a general principle of geometry that we will encounter again and again: In low dimensions, some conclusions become more difficult than for higher dimensions; geometry can develop its properties only when there is enough space.

If x, y are linearly dependent, we must use a representation $y = u + v$ for linearly independent u, v.

Question: Is F even *linear*, i.e. do we have $F(\lambda x) = \lambda F(x)$ for all $\lambda \in \mathbb{K}$? For each $x \neq 0$ the straight line $ox = \mathbb{K}x$ is bijectively mapped onto the straight line $F(o)F(x) = oF(x) = \mathbb{K}F(x)$. So for each $\lambda \in \mathbb{K}$ there is a $\bar{\lambda} \in \mathbb{K}$ with

$$F(\lambda x) = \bar{\lambda} F(x). \tag{2.8}$$

Let us consider a second vector y such that x and y are linearly independent. Then λy is the intersection of the line oy with the line parallel to xy through the point λx. Because F preserves parallelity, the image point $F(\lambda y)$ is characterized similarly, namely as the intersection of the straight line $oF(y)$ with the line parallel to $F(x)F(y)$ through the point $\bar{\lambda}F(x)$. According to the geometrical characterization of homotheties this is the point $\bar{\lambda}F(y)$, so we get for all $y \in X$:

$$F(\lambda y) = \bar{\lambda} F(y). \tag{2.9}$$

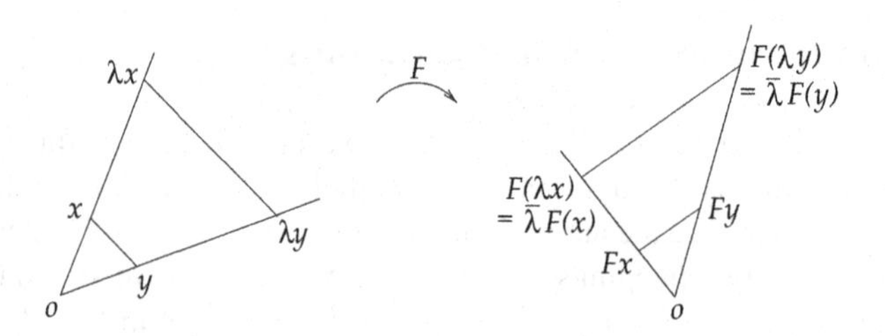

The scalar $\bar{\lambda}$ in (2.8) thus depends only on λ, not on x, i.e. $\lambda \mapsto \bar{\lambda}$ defines a bijective mapping $\mathbb{K} \to \mathbb{K}$.

We want to show that this mapping is a *field automorphism*, i.e.

$$\overline{\lambda + \mu} = \bar{\lambda} + \bar{\mu}, \tag{2.10}$$

$$\overline{\lambda \cdot \mu} = \bar{\lambda} \cdot \bar{\mu}. \tag{2.11}$$

Equation (2.10) follows with the additivity of F:

$$(\overline{\lambda + \mu})F(x) = F((\lambda + \mu)x) = F(\lambda x + \mu x) = F(\lambda x) + F(\mu x) = (\bar{\lambda} + \bar{\mu})F(x).$$

Equation (2.11) follows because we also have a geometric description of the multiplication $(\lambda, \mu) \mapsto \lambda\mu$: Given are points x, μx on the line ox as well as y and λy on another line oy, then $\lambda(\mu x)$ is the intersection of the line ox with the parallel to the line $\mu x \vee y$ (connecting the points μx and y) through the point λy. Since F maps this configuration into a corresponding one with $\bar{\lambda}$, $\bar{\mu}$ instead of λ, μ (see figure below), it follows

$$(\overline{\lambda \cdot \mu})F(x) = F((\lambda \cdot \mu)x) = (\overline{\lambda} \cdot \overline{\mu})F(x)$$

and hence Eq. (2.11).

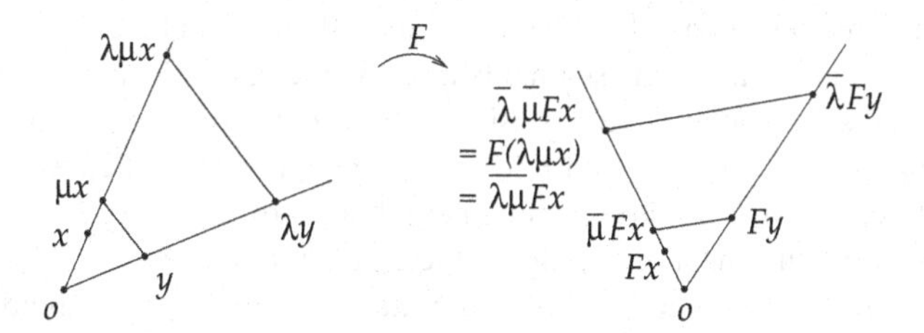

Such mappings F are called semilinear: If X, Y are two vector spaces over a field $\mathbb{K}$, a mapping $F : X \to Y$ is called *semilinear* if there is a field automorphism $\lambda \mapsto \overline{\lambda}$ on $\mathbb{K}$ with

$$F(x + y) = F(x) + F(y), \quad F(\lambda x) = \overline{\lambda} F(x). \tag{2.12}$$

Every linear mapping is in particular semilinear, because the identical mapping $\overline{\lambda} = \lambda$ is of course also a field automorphism on $\mathbb{K}$. If one extends F by an additive constant, one moves from semilinear to semiaffine mappings: $F : X \to Y$ is called *semi-affine* if there is a semilinear mapping $F_o : X \to Y$ and a translation T_a, $a \in Y$ such that $F = T_a F_o$, i.e.

$$F(x) = F_o(x) + a \tag{2.13}$$

for all $x \in X$. The following theorem is almost proved:

Theorem 2.1 *An invertible mapping $F : X \to X$ is parallel if and only if F is semi-affine.*

Proof A semilinear mapping F is parallel: If $L = \mathbb{K}x \subset X$ is a one-dimensional linear subspace, then $F(\lambda x) = \overline{\lambda} F(x) \in \mathbb{K}F(x)$, so $F(L) = F(\mathbb{K}x) = \mathbb{K}F(x) =: L'$ is again a one-dimensional subspace, and every straight line $L + y$ (being parallel to L) is mapped onto the straight line $F(L + y) = F(L) + F(y) = L' + F(y)$ which is parallel to L'. A translation is also parallel, so semi-affine mappings (the compositions of semilinear mappings with translations) are parallel too.

Conversely, we have already seen that a parallel map F_o with $F_o(o) = o$ is semilinear. If now F is an arbitrary parallel mapping with $F(o) = a$, then $F_o = T_{-a} F$ maps the point o back to itself, because $F_o(o) = F(o) - a = a - a = o$. So F_o is semilinear and hence $F = T_a F_o$ is semi-affine. □

Remark

1. In the case $\mathbb{K} = \mathbb{R}$, Theorem 2.1 has also a local version. It suffices to assume that F is defined on an open subset $U_1 \subset X = \mathbb{R}^n$ and is an invertibly continuous map[6] onto another open subset $U_2 \subset X$ such that straight line segments in U_1 are mapped onto straight line segments in U_2 while parallelism is preserved. The proof remains the same: By composing with a translation before and after, one can assume that $0 \in U_1 \cap U_2$ and $F(0) = 0$. The figures by which we have shown the semilinearity of F fit into the open neighborhoods U_1, U_2 of 0.
2. What are the semilinear mappings which are not already linear? To answer this question we only need to know the automorphisms of the field $\mathbb{K}$. Every automorphism of $\mathbb{K}$ fixes the distinguished elements 0 and 1, Hence all sums $1 + 1 + \ldots + 1$ and their additive and multiplicative inverses are fixed. Therefore the fields $\mathbb{K} = \mathbb{Q}$ and $\mathbb{K} = \mathbb{F}_p$ have no automorphisms except the identity. The field $\mathbb{R}$ has many automorphisms, but these must keep all rational numbers fixed. Therefore if we assume in addition that the automorphism is *continuous* (mapping limits to limits), then again only the identity remains. This is because $\mathbb{Q} \subset \mathbb{R}$ is *dense*, i.e. every real number is a limit of rational numbers. We will always tacitly make this assumption of continuity, so the notions of "semi-affine" and "affine" for $\mathbb{K} = \mathbb{R}$ are identical. For $\mathbb{K} = \mathbb{C}$ we already know a nontrivial automorphism: the conjugation, and there are no other continuous automorphisms either, since such an automorphism fixes every real number and maps i to a number $j \in \mathbb{C}$ with $j^2 = -1$, hence $j = \pm i$.[7]

2.4 Parallel Projections

We wish to extend results as in the previous section to mappings between *different* vector spaces X, Y of possibly different dimensions. Examples are *parallel projections* that often occur in drawings of spatial objects. For this we first have to modify the notion of parallel maps a little, because if F is no longer injective, a straight line might also be mapped onto a point. Therefore a mapping $F : X \to Y$ with $\dim X \geq 2$ (see Theorem 2.4 for $\dim X = 1$) will be called *parallel* if F maps every straight line either onto a point or bijectively onto a straight line, and thereby two parallel straight lines in X are mapped onto two points in Y or onto two (not necessarily different) parallel straight lines. First, we need a geometric characterization of subspaces:

[6] F is called invertibly continuous or *homeomorphism* if F is invertible and F as well as its inverse map F^{-1} are continuous.

[7] A *skew field* $\mathbb{K}$ (e.g. $\mathbb{K} = \mathbb{H}$) has a very large group of automorphisms, in particular all mappings $\lambda \mapsto \mu\lambda\mu^{-1}$ for fixed $\mu \neq 0$.

Lemma 2.2 *Let X be a vector space over a field $\mathbb{K}$ with $1 + 1 \neq 0$, that is* char($\mathbb{K}$) $\neq 2$. *A non-empty subset $U \subset X$ is an affine subspace if and only if for all $u, v \in U$ the straight line uv is entirely contained in U.*

Proof If $U \subset X$ is an affine subspace, that is $U = U_o + x$ for a linear subspace U_o, and if u, v are different points in U, then $u = x + u_o$ and $v = x + v_o$ for $u_o, v_o \in U_o$, and the straight line $uv = x + \mathbb{K}(u_o - v_o)$ lies entirely in U.

In order to prove the converse we may assume $o \in U$: If necessary we transform U to U_o by a translation. Thus by our assumption, for any $u \in U$ the straight line $ou = \mathbb{K}u$ is contained in U, that is $\lambda u \in U$ for all $\lambda \in \mathbb{K}$. Furthermore, for two different points $u, v \in U$ the straight line uv is contained in U. This line consists of the points of the form

$$v + \lambda(u - v) = \lambda u + (1 - \lambda)v, \ \lambda \in \mathbb{K},$$

and in particular $\frac{1}{2}(u + v) \in uv \in U$ and thus $u + v = 2 \cdot \frac{1}{2}(u + v) \in U$. So U a linear subspace. □

Remark For $\mathbb{K} = \mathbb{F}_2 = \{0, 1\}$ this characterization is wrong: The point set $\bar{U} = \{(0, 0), (1, 0), (0, 1)\} \subset \mathbb{K}^2$ satisfies the criterion, but is not a subspace, since $(1, 0) + (0, 1) = (1, 1) \notin \bar{U}$. More generally, the line uv contains only the points u and v, and $u + v$ is not a scalar multiple of a point on uv.

Theorem 2.3 *Let X, Y be vector spaces over a field $\mathbb{K}$ with* char($\mathbb{K}$) $\neq 2$ *and* $\dim X \geq 2$. *Then a map $F : X \to Y$ is parallel if and only if F is semi-affine.*[8]

Proof If F is semi-affine, i.e. $F(x) = S(x) + a$ for a semilinear mapping $S : X \to Y$, then S maps every one-dimensional subspace $L_o \subset X$ either to the null space or to a one-dimensional subspace $L'_o \subset Y$. So F maps two parallel straight lines $L_o + x$ and $L_o + x'$ onto the points $F(x)$ and $F(x')$ or onto the parallel lines $L'_o + F(x)$ and $L'_o + F(x')$, hence F is a parallel map.

Conversely, let F be a parallel map. By our lemma, $\mathrm{im}(F) \subset Y$ is an affine subspace, because with two distinct points $y_1 = F(x_1)$ and $y_2 = F(x_2)$, also the straight line $y_1y_2 = F(x_1x_2)$ is contained in $\mathrm{im}(F)$. By the same criterion, the full pre-image $F^{-1}(y)$ for each $y \in \mathrm{im}(F)$ is an affine subspace: If $x_1, x_2 \in F^{-1}(y)$ are different, then $F|_{x_1x_2}$ is not injective, since $F(x_1) = F(x_2) = y$; therefore, by definition of parallelity for maps, the image of the straight line x_1x_2 is the point y, and hence $x_1x_2 \subset F^{-1}(y)$. Also every subspace U' parallel to $U = F^{-1}(y)$ is mapped by F onto a point, because every straight line in U' through a fixed point $x' \in U'$ is parallel to a straight line in U and is therefore mapped likewise to a point, i.e. to $y' = F(x')$. Therefore $U' \subset F^{-1}(y')$ and in particular $\dim F^{-1}(y') \geq \dim F^{-1}(y)$. Interchanging the roles of y and y' we obtain the converse inequality

[8] Problem: Does the theorem also hold for $char(\mathbb{K}) = 2$?

and dimensional equality; thus $U' = F^{-1}(y')$ and therefore all pre-images are parallel affine subspaces.

Let us now choose a subspace $X_1 \subset X$ complementary to $U = F^{-1}(y)$ and set $Y_1 = \text{im}(F)$, then $F_1 = F|_{X_1} : X_1 \to Y_1$ is bijective (for X_1 intersects every $F^{-1}(y)$, and exactly once). Thus F_1 is semi-affine according to Theorem 2.1, because we can identify X_1 and Y_1 by a linear isomorphism. We may assume (after shifting if necessary) that X_1 and U pass through the origin 0 and therefore are linear subspaces and $X = X_1 \oplus U$. Let us denote by $p_1 : X \to X_1$ the projection onto the direct summand X_1 and by $i_1 : Y_1 \to Y$ the inclusion mapping, then $F = i_1 F_1 p_1$. Thus the mapping is itself semi-affine as a composition of semi-affine mappings. □

The best known examples of such mappings are the *parallel projections,* the most common form of two-dimensional drawings of three-dimensional objects in mathematics, science, and engineering.

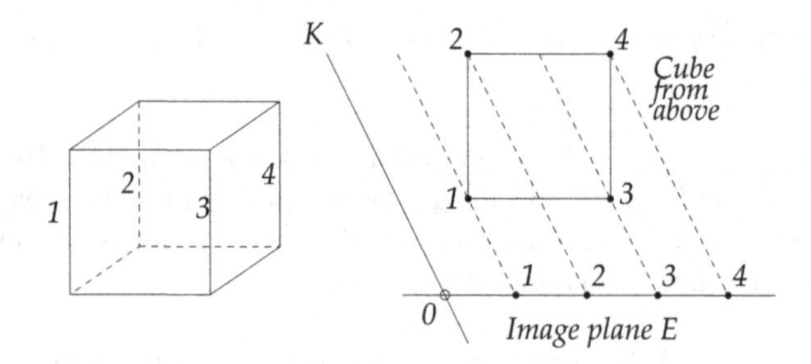

For this, one decomposes the space $\mathbb{R}^3$ into the image plane E and an arbitrary (oblique) one-dimensional vector space complement K, i.e. $\mathbb{R}^3 = K \oplus E$. Each $x \in \mathbb{R}^3$ thus possesses a unique decomposition $x = k + e$ with $k \in K$ and $e \in E$. The *projection onto E along K* is the mapping $F : \mathbb{R}^3 \to E$, $k + e \mapsto e$: The K-component is suppressed, and the projection is linear with kernel K. The straight lines that are mapped to points are called *projection lines*. These are exactly the parallels to K, and their image points are their intersections with the image plane E.

2.5 Affine Representations, Ratio, Center of Gravity

Let X be an n-dimensional affine space over $\mathbb{K}$. An *affine basis* of X is an $(n+1)$-tuple of *points* $a_0, a_1, \dots, a_n \in X$ with the property that the *vectors* $a_1 - a_0, \dots, a_n - a_0$ are linearly independent;[9] such points are also called *affinely*

[9] We keep the language as if X were a vector space. Actually, yes, there is a vector space V distinct from X whose additive group acts simply transitively on X by means of an action $(v, x) \mapsto T_v x$, that is: For any $a_0, x \in X$ there is a uniquely determined vector $v \in V$ with $T_v a_0 = x$. But to keep notation intuitive we still write $v =: x - a_0$ and $T_v a_0 =: v + a_0$. The distinction between X and

independent. Then every point $x \in X$ can be uniquely represented as

$$x = \sum_{j=0}^{n} \lambda_j a_j \text{ with } \lambda_j \in \mathbb{K} \text{ and } \sum_{j=0}^{n} \lambda_j = 1. \tag{2.14}$$

Because the vectors $b_i = a_i - a_0$ form a vector space basis, the vector $x - a_0$ has a representation $x - a_0 = \sum_{i=1}^{n} \lambda_i b_i = \sum_{i=1}^{n} \lambda_i a_i - (\sum_{i=1}^{n} \lambda_i) a_0$ and thus $x = \sum_{j=0}^{n} \lambda_j a_j$ with $\lambda_0 = 1 - \sum_{i=1}^{n} \lambda_i$. The numbers $\lambda_0, \dots, \lambda_n$ are called the *affine coordinates* for x with respect to the affine basis $a_0, \dots, a_n$.

What is interesting about this representation is its invariance under affine mappings: If Y is a second affine space over $\mathbb{K}$ and $F : X \to Y$ an affine mapping, that is $F(x) = F_o(x) + b$ for a linear mapping F_o and a fixed b, then these numbers are preserved: For $x = \sum_j \lambda_j a_j$ with $\sum_j \lambda_j = 1$ we have

$$\begin{aligned} F(x) &= \sum_j \lambda_j F_o(a_j) + b \\ &= \sum_j \lambda_j F_o(a_j) + \sum_j \lambda_j b \\ &= \sum_j \lambda_j (F_o(a_j) + b) \\ &= \sum_j \lambda_j F(a_j). \end{aligned}$$

The point $F(x)$ thus has the same position with respect to the points $F(a_j)$ as the point x with respect to a_j.

Linear combinations of the form (2.14) still make sense and are still invariant under affine maps if the generating points a_j no longer form an affine basis and their number is arbitrary, say $r + 1$, but this representation of x need not be unique. An important special case is the *center of gravity* or the *arithmetic means* for which all λ_j are equal, so $\lambda_j = \frac{1}{r+1}$ for $j = 0, \dots, r$. Clearly, the center of gravity is preserved even by semi-affine mappings.

Another special case is $r = 1$: The points x on the straight line $a_0 a_1$ are parametrized by the number $\lambda \in \mathbb{K}$ with

$$x = \lambda a_1 + (1 - \lambda) a_0 = a_0 + \lambda(a_1 - a_0) \quad \text{or} \quad x - a_0 = \lambda(a_1 - a_0).$$

V is emphasized only by the labels “points” and “vectors”: *Points* are elements of X, while *vectors* are elements of V.

It gives the *ratio* $\lambda =: \frac{x-a_0}{a_1-a_0} = r(x, a_1, a_0)$ of the linearly dependent vectors $x - a_0$ and $a_1 - a_0$,[10] and thus the relative position of the point x with respect to a_0 and a_1 is given. The ratio makes possible a geometric characterization of affine mappings even on a *one-dimensional* affine space:

Theorem 2.4 *Let X be a one-dimensional affine space. A bijection $F : X \to X$ is affine if and only if F preserves the ratio: $r(Fx, Fy, Fz) = r(x, y, z)$ for all $x, y, z \in X$.*

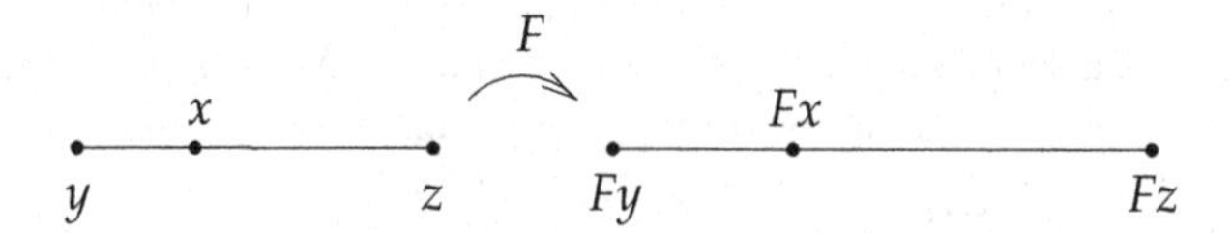

Proof Affine mappings, in particular translations, obviously have this property. Conversely, suppose that F preserves the ratio. By postcomposing with a translation, we may assume $F(0) = 0$. For any $\lambda \in \mathbb{K}$ and $y \in X \setminus \{0\}$ let $x = \lambda y \in X$. Then $\lambda = r(x, y, 0)$. By assumption, $r(F(x), F(y), 0) = r(x, y, 0) = \lambda$. Thus $F(x) = \lambda F(y)$ for all $\lambda \in \mathbb{K}$ which in dimension one also implies additivity: $F(x + y) = F(\lambda y + y) = F((\lambda + 1)y) = (\lambda + 1)F(y) = \lambda F(y) + F(y) = F(x) + F(y)$. Therefore F is linear. □

Remark Clearly, a bijective map $F : X \to X$ is semi-affine on a one-dimensional affine space X over $\mathbb{K}$ if and only if there is an automorphism σ of $\mathbb{K}$ such that $f(Fx, Fy, Fz) = \sigma(r(x, y, z))$ for all $x, y, z \in X$.

---

[10] Note that $x - a_0$ and $a_1 - a_0$ are not numbers, but vectors; however their "quotient" (ratio) is a number. The same is true for *quantities* (like lengths, volumes, masses) which was already known to Pythagoras. In modern terminology, scalar quantities of the same kind form a one-dimensional vector space over $\mathbb{R}$, and a *unit* (meter, liter, kilogram) is a basis vector. For the notion of ratio and its importance for the discovery of irrational numbers see also the first chapter in "Sternstunden der Mathematik" [12].

3 Incidence: Projective Geometry

Abstract

Projective geometry is the proper domain for the notion of incidence for straight lines and points; it knows no other basic notions. The basic ideas were developed 600 years ago with the discovery of central perspective. The artists pioneered the mathematicians. Just as in a perspective view parallel lines appear to have a point of intersection on the horizon, so parallelism is interpreted as "intersecting at infinity". For this, geometry must be extended by "points at infinity". These arise quite easily by embedding into linear algebra: This extended ("projective") geometry, however, no longer takes place on a vector space as affine geometry does, but on the set of its one-dimensional linear subspaces. The structure-preserving transformations ("collineations") are then simply the (semi-)linear isomorphisms of the vector space. Now for the first time in this book interesting geometric theorems are discussed, the theorems of *Desargues, Brianchon* and *Pascal.* We will get to know conic sections and quadrics, and at the end an important numerical quantity which is invariant under projective transformations: the cross-ratio.

3.1 Central Perspective

Are there mappings in the intuitive plane or space which map straight lines into straight lines (i.e. preserving incidence) without being parallel (affine)? Such maps are well known to us from photographs: perspective pictures. To construct central perspective correctly, one needs only three simple rules:

1. Straight lines are mapped into straight lines,
2. images of parallels are parallel again or have a common intersection point (*vanishing point*),

J.-H. Eschenburg, *Geometry – Intuition and Concepts*,
https://doi.org/10.1007/978-3-658-38640-5_3

3. images of parallels to straight lines in a fixed plane intersect on a common straight line, the *horizon* of that plane .

The simplest exercise in perspective drawing is a railroad track that runs straight toward the horizon and whose railroad ties are evenly spaced. Then one only needs to specify the horizon, the two tracks and the first two crossties in the picture. The images of the other crossties can be constructed, for all the rectangles formed by the tracks and two adjacent crossties have parallel diagonals whose images (if straightly extended)[1] intersect in a common point on the horizon. (It is not essential for the images of the crossties to be parallel.)

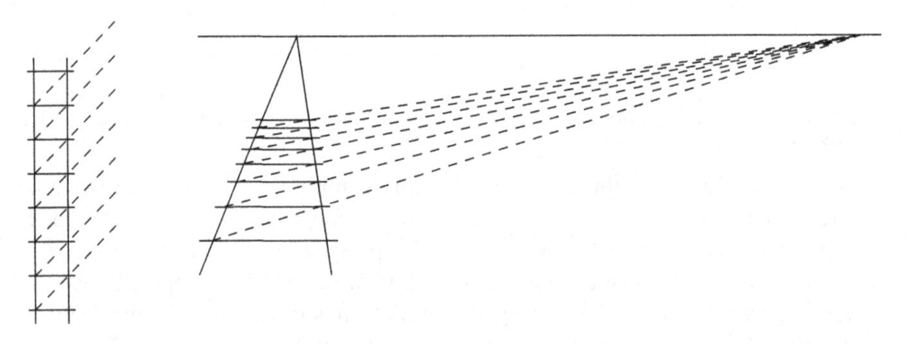

If we replace the railway tracks and the crossties by the straight lines of a rectangular coordinate system in the plane, we see the same pattern: Under a perspective mapping, the image of a single rectangle, especially a coordinate box, can be any convex planar quadrilateral, and it uniquely determines the image of every other point in the plane.

The same principles apply to drawings of *spatial* objects, such as a cuboid. The vertical edges are usually drawn vertically as well (so this family of parallel straight lines is mapped to parallels). If we specify the image of the front rectangle and one of the side rectangles, everything else is determined.

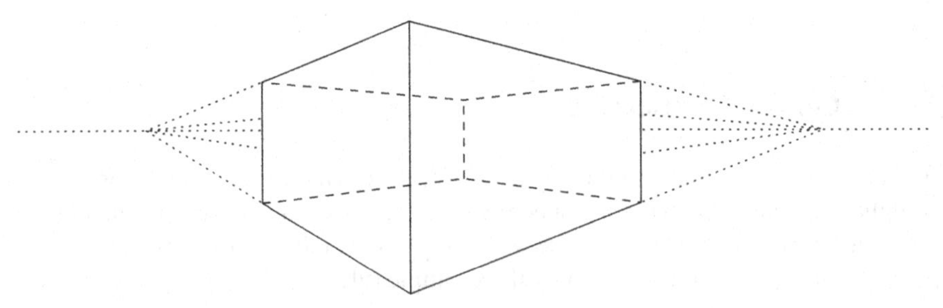

By putting on a roof, we get the image of a gabled house; we only have to specify the height of the front gable, cf. Exercise 11.

[1] These straight extensions are called *vanishing lines*.

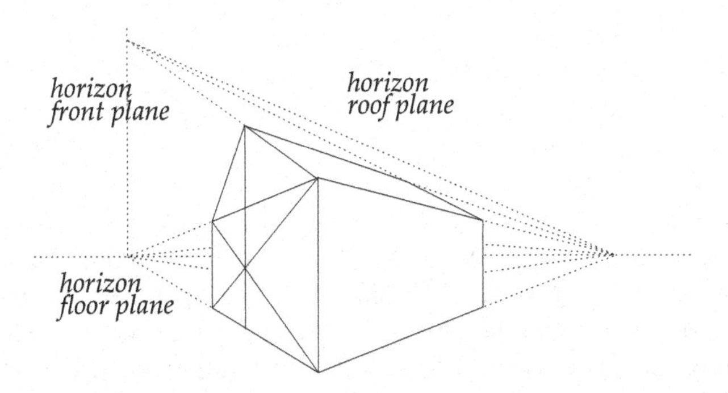

Today we are familiar with perspective from photography, but the people of earlier centuries had no such facility. Our vision is not really perspective, because our brain receives additional depth information through the binocularity and the adaptation of the eye lens to the distance. Perspectively, an object should seem to get larger as we approach it, but at close range this is not the case at all; the object seems to maintain its size. Perspective representation thus presupposes a certain abstraction of natural vision. It is a discovery of the early Renaissance, probably the first significant mathematical contribution of Europe since antiquity. There had been attempts from antiquity to render spatial depth by oblique and convergent lines, but the precise construction remained obscure. It was not until around 1420 that the later master builder of the dome of Florence Cathedral succeeded, *Filippo Brunelleschi* (1377–1440), the drawings of which, however, we know only from reports. The first perspective pictures that have come down to us were made by a friend of Brunelleschi, the painter *Masaccio* (actually Tomaso di Giovanni di Simone, 1401–1428). Particularly famous is his fresco "Trinity" (1426) in the church of Santa Maria Novella in Florence,[2] in which perspective takes on an important function in the painting's message because it involves the viewer's position. The first textbook on perspective was written by the Genoese scholar *Leon Battista Alberti* (1404–1472).

In affine geometry we have discussed in Sect. 2.4 the *parallel projections*. Perspective illustrations, on the other hand, are *central projections*. Here, too, the image point is created as the intersection of the image plane with a straight line passing through the pre-image point, the *projection line*. However, the projection lines are no longer parallel, but instead they all pass through a fixed point called *projection center*.[3] In perspective vision the eye itself is the center of projection; the straight lines of projection are the rays of light which, starting from the object A, reach the eye, and the image point A' is the intersection of this ray with the image plane, which is conceived to be between the eye and the objects.

[2] https://en.wikipedia.org/wiki/Holy_Trinity_(Masaccio).

[3] The parallel projections in Sect. 2.4 are a special case of central projections, where the center of projection is in the "plane at infinity", see Sect. 3.2. The affine space $\mathbb{R}^3$ must be extended to the projective space $\mathbb{RP}^3$, see below.

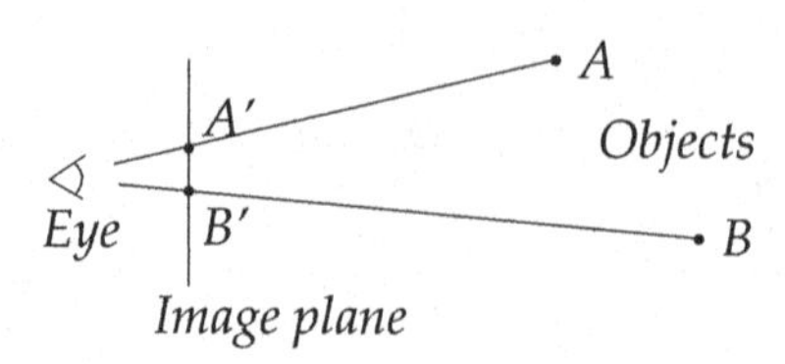

Albrecht Dürer (1471–1528, Nuremberg) shows in his textbook "Underweysung der Messung" of 1525 how to determine the image point on a pane of glass placed between the eye and the object by aiming at the object through a hole in a permanently mounted frame.[4] He also describes a purely mechanical method of producing a perspective image, in which the projection rays are replaced by threads:[5]

> "If thou be in a hall, strike a great needle with a wide eye made for that purpose into a wall, and set that for an eye. Through that draw a strong thread, and hang a lead weight at the bottom of it. Then set a table or board as far from the eye of the needle as the thread will be in it. Set thou an upright frame upon it ... with a little door that may be opened and shut. This little door is your board, on which you want to paint. Then nail two strings as long as the upright frame is long and wide into the top and middle of the frame, and the other on one side also into the middle of the frame, and let them hang. Then make a long pin, which has an eye of a needle at the front. Thread into it the long thread which is drawn through the eye of the needle on the wall, and go out through the frame with the needle and the long thread, and give it into the hand of another, and wait thou of the other two threads which hang on the frame. Now use this: Lay a lute, or whatever else pleases you, as far from the frame as you like, and that it may remain unmoved as long as you need it, and let your journeyman stretch out the needle with the thread on the most necessary points of the lute, and as often as he stops on one place and stretches out the long thread, then strike the two threads on the frame crosswise stretched to the long thread and stick them in both places with a wax to the frame, and let your journeyman slacken his long thread. Then shut the little door, and draw the same points, where the threads go crosswise over each other, on the board. Then open the little door again and do with another point, but so until you dot the whole lute on the board. Then draw all the dots that have been on the board from the sound with lines together, so you see what becomes of it. So mayst thou mark other things also."

In the case of the camera or its predecessor, the pinhole camera (camera obscura), things are somewhat different: The projection center is the lens center or hole, and the image plane is located behind it on the back wall of the camera:

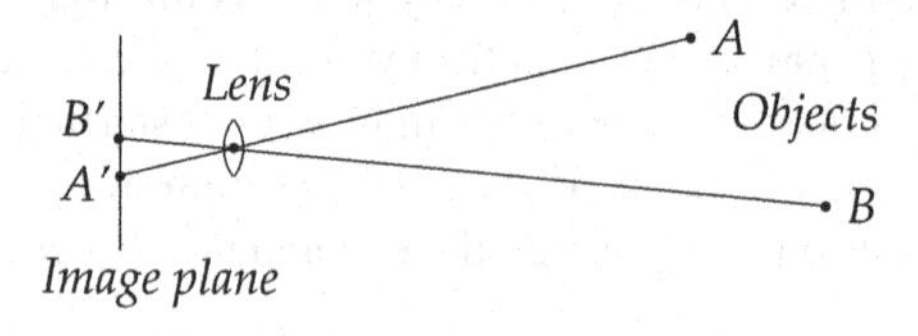

[4] https://de.wikisource.org/wiki/Seite:Duerer_Underweysung_der_Messung_180.jpg.

[5] https://de.wikisource.org/wiki/Seite:Duerer_Underweysung_der_Messung_181.jpg.

So the image plane is no longer between the object and the projection center which is now the lens, but it is behind the projection center. However, this difference is not essential: A parallel shift of the image plane causes just a homothety S_λ of the image. If the image plane is shifted to the other side of the center of projection, as in the present case, then λ is negative. The image is therefore rotated by 180° in the camera, i.e. it is upside down.

3.2 Points at Infinity and Projection Lines

The French fortress builder *Gerard Desargues* (1591–1661) developed an idea which was to prove very far-reaching. In a perspective image of a plane there is a straight line on which the images of parallel straight lines meet, the horizon. But no straight line of the depicted plane corresponds to it. Shouldn't the original image plane be extended by new points "lying at infinity", so-called *points at infinity* or *ideal* points, i.e. points existing only in idea, which could be regarded as pre-images of the horizon points? The points at infinity would have to form together a new straight line, the *line at infinity*, the pre-image of the horizon. Then one would finally get rid of the annoying special case of affine geometry, that two straight lines of a plane unfortunately do not always have an intersection point, but are sometimes parallel: The parallel straight lines would just meet in the newly gained points, the points at infinity, and to each class of parallel straight lines exactly one such point at infinity would belong.[6] In the same way one could speak in space of an (imaginary) *plane at infinity* which contains the intersections of parallel lines and which contains the lines at infinity for all planes in space. The fact that such points do not really exist did not bother mathematicians very much; it was just an extension of the usual affine geometry by new, "ideal" points, similar to the way numbers could be extended by adding imaginary new numbers (e.g. ∞). This extension was called *projective geometry*. When the French mathematician *J.-V. Poncelet*[7] became a prisoner of war in Russia in 1812 and had a lot of time but no books at his disposal, he systematically developed the laws of this geometry.

The definition of the points at infinity as classes of parallel lines has one drawback: Points at infinity and ordinary points seem to be defined quite differently. Is there no *common* definition? Again, the perspective mappings (the central projections as in photography) give the key. We describe them once more in the terms of spatial affine geometry. Every point x of the original plane U determines exactly one straight line ox through the projection center o, and its image point is the intersection of this straight line with the image plane B. Actually we can forget the points x of the original plane and replace them by their projection lines ox.

[6] One can define a *point at infinity* in exactly this way: as a class of parallel straight lines, that is, as an equivalence class of a straight line of the affine plane with respect to the equivalence relation "parallelism" on the set of straight lines.

[7] Jean-Victor Poncelet, 1788 (Metz)–1867 (Paris).

Projection lines whose intersections with B lie on a common straight line in B are contained in a common plane, namely the plane spanned by that line in B and the projection center. We have therefore found a kind of dictionary: Points correspond to projection lines through o and lines correspond to planes through o.

But some straight lines through o do not meet the original plane U at all, namely those which are parallel to a straight line within U. These do not correspond to any point of U, but they are the "projection lines" of the new "ideal" points of U. According to our dictionary, these points lie on a common "straight line" (the line at infinity), for their projection lines are all in the plane through o parallel to U.[8] On the other hand, these straight lines may very well intersect the image plane B, hence we see the horizon there as the "image of the line at infinity" (cf. Exercise 12).

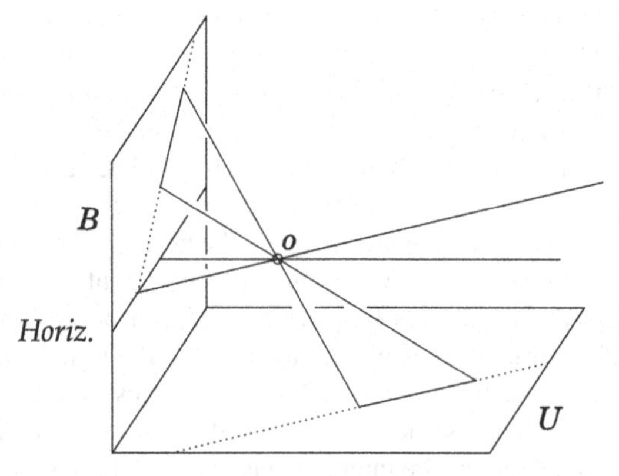

Planar projective geometry is therefore nothing but the geometry of the "bundle" of straight lines through a fixed point o in space, using only new words: A straight line through o will be called "point" and a plane through o "straight line". The whole bundle stands for the projective plane, the straight lines of the bundle which meet the plane U represent the affine plane, and those that do not meet U represent the line at infinity.

If we define the affine space with the distinguished point o again as a vector space V with origin o, then this bundle of straight lines is just the *set of all one-dimensional linear subspaces of V*.

3.3 Projective and Affine Space

In general, let us now consider an arbitrary vector space V over a field $\mathbb{K}$. We define the projective space over V as follows:

Definition: The *projective space* P_V over V is the set of all one-dimensional linear subspaces of V. For $V = \mathbb{K}^{n+1}$ we denote P_V also by $\mathbb{KP}^n$ or simply $\mathbb{P}^n$.

[8] Each such straight line defines a class of straight lines parallel to it in U; this is the associated point at infinity in the sense of the previous definition.

We say that P_V has dimension n if $\dim V = n + 1$. Any $(k + 1)$-dimensional linear subspace $W \subset V$ corresponds to a *k-dimensional projective subspace* $P_W \subset P_V$ whose elements are the one-dimensional linear subspaces of W. In particular, a *straight line* in P_V consists of the one-dimensional linear subspaces of a two-dimensional linear subspace of V.[9]

One can also describe P_V as follows: Two vectors $v, w \in V_* := V \setminus \{0\}$ are called *proportional*, $v \sim w$, if there is a number $\lambda \in \mathbb{K}^*$ with $w = \lambda v$. This is obviously an equivalence relation, and it is related, as in Exercise 2, to a group action S, namely with the action S of the group $\mathbb{K}^*$ on V through multiplication by scalars, $S : \mathbb{K}^* \times V \to V$, $S(\lambda, v) = S_\lambda(v) = \lambda v$. The orbit of a vector $v \in V_*$ under this group action, the equivalence class $[v]$, is the one-dimensional subspace generated by v (intersected with V_*, i.e. without the origin) and thus an element of P_V. We therefore obtain

$$P_V = \{[v] = K^* v;\ v \in V_*\}. \tag{3.1}$$

The equivalence class $[v]$ is the vector v "up to multiples"; one calls $[v]$ also a *homogeneous vector.* By $\pi : V_* \to P_V$, $\pi(v) = [v]$ we denote the canonical projection.

How does P_V extend affine space? As affine space we consider any affine hyperplane $H \subset V$ which does not pass through the origin:

$$H = W + v_o\,,$$

where $W \subset V$ is a linear subspace of codimension one with $v_o \notin W$. Most one-dimensional linear subspaces in V intersect H (and exactly once), only those parallel to H (that is: contained in W) do not intersect. They form the projective hyperplane P_W, which we want to call the *hyperplane at infinity* belonging to H. The remaining points form the subset

$$A_H = \pi(H) = [H] = \{[v] \in P_V;\ v \in H\} \subset P_V\,, \tag{3.2}$$

which we call the *affine space* in P_V; in fact $\pi|_H : H \to A_H$ is bijective and *straight*, i.e., straight lines in H (intersections of H with a two-dimensional transversal[10] subspace E) are mapped to projective straight lines as far as they are in A_H, and vice versa. Thus the projective space P_V is a disjoint union of the affine space A_H and the hyperplane at infinity P_W:

$$P_V = A_H \,\dot\cup\, P_W. \tag{3.3}$$

---

[9] Similarly, for any dimension k between 1 and $\dim V$ the set of all k-dimensional linear subspaces W of V can be considered. This set is called *Grassmann manifold* $G_k(V)$. In particular $G_1(V) = P_V$. Thanks to the correspondence $W \mapsto P_W$ one can also regard $G_k(V)$ as the set of $(k-1)$-dimensional projective subspaces of P_V.

[10] Two subspaces of V are called *transversal* if their union spans the whole space V.

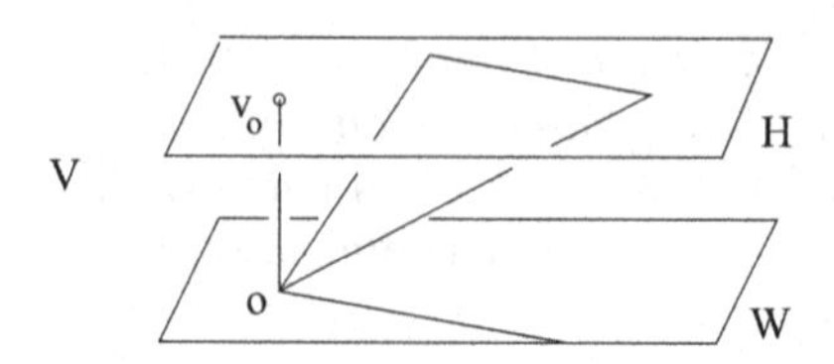

Theorem 3.1 *Projective straight lines $g_1, g_2 \subset P_V$ intersect at a point of P_W, the hyperplane at infinity, if and only if $g_1 \cap A_H$ and $g_2 \cap A_H$ are parallel straight lines in A_H, i.e. images under $\pi|_H$ of parallel straight lines in H.*

Proof Let $\tilde{g}_1, \tilde{g}_2$ be parallel straight lines in H, so $\tilde{g}_i = L + v_i$ for a one-dimensional linear subspace $L = \mathbb{K}v \subset W$ and $v_i \in H$. Then $\tilde{g}_i = H \cap E_i$, where E_i is the two-dimensional subspace spanned by v and v_i, and $\pi(\tilde{g}_i) = \pi(E_i) =: g_i$. Since $E_1 \cap E_2 = L$ we have $g_1 \cap g_2 = \pi(L) \in P_W$.

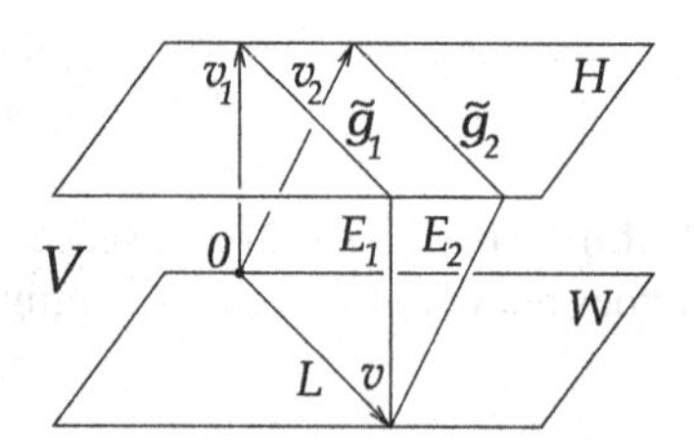

Conversely, let $g_1 = \pi(E_1)$ and $g_2 = \pi(E_2)$ be straight lines in P_V with an intersection $[v] \in P_W$, and $L = \mathbb{K}v \subset W$ be the one-dimensional linear subspace corresponding to $[v]$. Then $L = E_1 \cap E_2$. The plane E_i is spanned by v and a vector $v_i \notin W$, and a multiple of v_i meets the hyperplane H parallel to W, because W and v_i generate V. So we can assume $v_i \in H$, and $H \cap E_i = L + v_i =: \tilde{g}_i$. These are parallel straight lines in H. □

We will often fix the vector space V as $\mathbb{K}^{n+1}$ and put $H = \mathbb{K}^n + e_{n+1}$. The projective space over $\mathbb{K}^{n+1}$ is denoted by $\mathbb{P}^n$ and $\mathbb{A}^n$ denotes the affine space $[H] = \{[x_1, \dots, x_n, 1]; x_1, \dots, x_n \in \mathbb{K}\}$ contained in $\mathbb{P}^n$. From (3.3) we thus obtain the decomposition

$$\mathbb{P}^n = \mathbb{A}^n \,\dot{\cup}\, \mathbb{P}^{n-1}. \tag{3.4}$$

This is no restriction since the general case can be transformed into this special case by a linear isomorphism between V and $\mathbb{K}^{n+1}$ (cf. Sect. 3.4).

The real projective plane as the bundle of all straight lines through the origin 0 in 3-space $\mathbb{R}^3$ can be imagined quite well: Let us consider the *sphere* $\mathbb{S}$ centered at the origin. Every straight line intersects this surface in two opposite *(antipodal)* points

$\pm v$, $v \in \mathbb{S}$. Hence we can thus think of the projective plane as the set of antipodal pairs of points of the sphere,

$$\mathbb{P}^2 = \mathbb{S}/\pm\,. \tag{3.5}$$

In other words, we obtain the projective plane by gluing the northern hemisphere to the southern hemisphere in such a way that just the antipodal points are stitched together. If we consider from the sphere at first only a band around the equator, this gluing can be practically done; the result is the *Möbius strip*, a closed band with a twist of half a turn (180°).[11] After this, what remains of the sphere are the two polar caps, which we can glue together antipodally to a single cap. This cap must now be glued again to the Möbius strip, which, like the cap, is bordered by a single closed line. This gluing is not so easy to do in space, but abstractly it is no problem.[12] So the projective plane is the union of a Möbius strip and a disk whose borders are glued together.

With the sphere model we can also understand the *geometry* of the projective plane. Projective straight lines correspond to planes through 0, and these intersect the sphere in *great circles* which are thus corresponding to projective straight lines. Two great circles intersect each other in an antipodal pair of points which corresponds to the intersection point of the lines in projective plane. The *line at infinity* is the equator, the intersection of $\mathbb{S}$ with the equatorial plane $\mathbb{R}^2 = \{x \in \mathbb{R}^3;\ x_3 = 0\}$, with antipodal points glued together. The affine plane $\mathbb{A}^2 \subset \mathbb{P}^2$ can be imagined as the open upper hemisphere $\mathbb{S}_+ = \{x \in \mathbb{S};\ x_3 > 0\}$. Two great circle arcs in $\mathbb{S}_+$ intersect exactly once or share an antipodal pair of points on the equator. The latter case corresponds to a parallel pair of straight lines.

3.4 Semiprojective Mappings and Collineations

We still consider a vector space V over $\mathbb{K}$ and the corresponding projective space $P = P_V$. We want to know all *straight* invertible mappings $F : P \to P$ where "straight" means that lines are mapped onto lines. We will briefly call them *collineations*. (In the case $\mathbb{K} = \mathbb{R}$ or more generally $\mathbb{K} \supset \mathbb{R}$ we will also demand *continuity* of F and F^{-1}.)

[11] The Möbius strip is a unilateral ("non-orientable") surface; you cannot paint it green on one side and red on the other.

[12] One considers the disjoint union of the Möbius strip M with the cap C, maps the boundary of M onto the boundary of C with a homeomorphism f, and considers points on the boundary of M as identical to their image on the boundary of C. This is done with an equivalence relation on $M \dot\cup C$, according to which a point is equivalent only to itself or, if appropriate, to its image or pre-image under f. The resulting surface can be realized in a variety of ways as a surface in space (called *Boy surface*) when we allow self-intersections. A particularly beautiful example of a Boy surface is located in front of the Mathematical Research Institute Oberwolfach, see https://upload.wikimedia.org/wikipedia/commons/b/b1/Boyflaeche.JPG.

Definition: A *collineation of* P_V is an invertible mapping $F : P_V \to P_V$ with the property that $F(g)$ for any straight line $g \subset P_V$ is again a straight line in P_V.

The collineations obviously form a *group:* Compositions and inverses of collineations are collineations again. For compositions this is clear, and the inverse F^{-1} transforms the straight line $F(g)$ onto the straight line g, so it is also a collineation.

Examples of collineations on P_V are the invertible semilinear mappings on V, since they map linear subspaces onto linear subspaces of the same dimension; in particular, they preserve the set of straight lines and planes through the origin. Any invertible semilinear mapping $S : V \to V$ thus defines a collineation $F = [S] : P_V \to P_V$,

$$[S][v] = [Sv] \tag{3.6}$$

for all $v \in V_*$. Let us call such mappings *semiprojective mappings* and if S is linear (not just semilinear), they are called *projective mappings*. The semiprojective mappings at first sight form a subgroup of the group of collineations; but we shall show that this already exhausts all collineations.

Remark The group of invertible linear mappings on $\mathbb{K}^{n+1}$ is called $GL(\mathbb{K}^{n+1})$ ("General Linear Group"). It operates on $\mathbb{KP}^n$ by (3.6). The action of this group on $\mathbb{KP}^n$ has, however, a *kernel*, a subgroup whose elements act as the identical map on $\mathbb{RP}^n$. If $\mathbb{K}$ is commutative, these are exactly the multiples of the unit matrix (cf. Exercise 16); they form the group $\mathbb{K}^* = \{tI;\ t \in \mathbb{K}^*\}$, the center[13] of $GL(\mathbb{K}^{n+1})$. The effective group is the quotient group $PGL(\mathbb{K}^{n+1}) := GL(\mathbb{K}^{n+1})/\mathbb{K}^*$, called the *projective group*.

We again think of the affine space as a subset of the projective space, by removing a hyperplane $H = W + v_o$ with $v_o \notin W$ and $A_H := [H]$ where $W \subset V$ is a linear subspace of codimension one. In the case $V = \mathbb{K}^{n+1}$ (in this case we write $\mathbb{P}^n$ instead of P_V) one likes to choose $W = \mathbb{K}^n = \{(x_1, \dots, x_{n+1}) \in \mathbb{K}^{n+1};\ x_{n+1} = 0\}$ and $v_o = e_{n+1} = (0, \dots, 0, 1)$, so $H = \mathbb{K}^n + e_{n+1}$ and

$$A_H = \mathbb{A}^n = \{[x, 1];\ x \in \mathbb{K}^n\} \cong \mathbb{K}^n. \tag{3.7}$$

This is not a restriction, because every hyperplane $H \not\ni 0$ can be transformed by a linear isomorphism of $\mathbb{K}^{n+1}$ into the special hyperplane $\mathbb{K}^n + e_{n+1}$.

This embedding of the affine into the projective space gives us the following natural extension of each semi-affine map F on $\mathbb{K}^n \cong \mathbb{A}^n \subset \mathbb{P}^n$ to a semiprojective mapping $\hat{F}$ on $\mathbb{P}^n$: Is $F(x) = F_o(x) + a$ for a constant vector $a \in \mathbb{K}^n$ and

[13] The *center* of a group G consists of all elements $z \in G$ that interchange with any element of G, $zg = gz$ for all $g \in G$.

an invertible semilinear mapping F_o on $\mathbb{K}^n$ (with $F_o(\lambda x) = \bar{\lambda} F_o(x)$ for some automorphism $\lambda \mapsto \bar{\lambda}$ from $\mathbb{K}$), we obtain on $\mathbb{A}^n \subset \mathbb{P}^n$ the assignment

$$\begin{aligned} [x, 1] &\mapsto [F(x), 1], \\ [x, \xi] = [\xi^{-1}x, 1] &\mapsto [F(\xi^{-1}x), 1] = [\bar{\xi}^{-1}F_o(x) + a, 1] = [F_o(x) + \bar{\xi}a, \bar{\xi}] \end{aligned}$$

for all $\xi \in \mathbb{K}^*$. But the last assignment rule

$$[x, \xi] \mapsto [F_o(x) + \bar{\xi}a, \bar{\xi}]$$

is also defined in the case $\xi = 0$. Therefore, we can define the following invertible semilinear mapping $\hat{F}_o$ on $\mathbb{K}^{n+1}$:

$$\hat{F}_o(x, \xi) = (F_o(x) + \bar{\xi}a, \bar{\xi}) \text{ for all } x \in \mathbb{K}^n, \ \xi \in \mathbb{K}, \tag{3.8}$$

and the corresponding semiprojective mapping $\hat{F} = [\hat{F}_o]$ on $\mathbb{P}^n$ is the projective map which restricts to F on $\mathbb{A}^n$. When F_o is even linear, the same holds for $\hat{F}_o$, and we can use matrix notation with respect to the decomposition $\mathbb{K}^{n+1} = \mathbb{K}^n \oplus \mathbb{K}$:

$$\hat{F}_o = \begin{pmatrix} F_o & a \\ 0 & 1 \end{pmatrix}.$$

Theorem 3.2 *The collineations of $\mathbb{P}^n$ are exactly the semi-projective mappings: For each collineation $\hat{F}$ of $\mathbb{P}^n$ there is an invertible semilinear mapping $\hat{F}_o$ to $V = \mathbb{K}^{n+1}$ with $\hat{F} = [\hat{F}_o]$.*

Proof We trace the assertion back to the corresponding theorem on affine geometry (Theorem 2.1), by first considering only those collineations $\hat{F}$ which leave invariant the affine space $\mathbb{A}^n \subset \mathbb{P}^n$, $\hat{F}(\mathbb{A}^n) = \mathbb{A}^n$. Such bijective mappings also leave invariant the hyperplane at infinity $\mathbb{P}^{n-1} = \mathbb{P}^n \setminus \mathbb{A}^n$. Thus $F := \hat{F}|_{\mathbb{A}^n}$ does not only map lines to lines, but also parallels to parallels, because parallels in $\mathbb{A}^n$ are exactly the pairs of straight lines which intersect in a point of the hyperplane at infinity $\mathbb{P}^{n-1}$, and $\hat{F}$ preserves $\mathbb{P}^{n-1}$. In Theorem 2.1 we have shown that such a mapping F is semi-affine: $F(x) = F_o(x) + a$ for all $x \in \mathbb{K}^n \cong \mathbb{A}^n$. This semi-affine mapping can be continued, as shown above, to a semiprojective mapping, which coincides with $\hat{F}$ on $\mathbb{A}^n$ and thus must be everywhere equal to $\hat{F}$ because the hyperplane at infinity $\mathbb{P}^n \setminus \mathbb{A}^n$ consists of the intersections of "parallel" straight lines in $\mathbb{A}^n$. So $\hat{F}$ is semi-projective.

We trace the general case back to the special case just discussed. Any collineation $\hat{F}$ on $\mathbb{P}^n$ has the following property: It maps not only straight lines onto straight lines, but also k-dimensional projective subspaces onto k-dimensional projective subspaces for any $k \leq n$, as is easily shown by induction over k (Exercise 18). In particular, the hyperplane at infinity $\mathbb{P}^{n-1} = [\mathbb{K}^n]$ is mapped onto a projective hyperplane $[W] \subset \mathbb{P}^n$ (where $W \subset \mathbb{K}^{n+1}$ denotes a linear hyperplane, a linear subspace of codimension one). Now we choose an invertible linear map A on $\mathbb{K}^{n+1}$ with $A(\mathbb{K}^n) = W$; the corresponding projective map $\hat{A} = [A]$ then transforms $\mathbb{P}^{n-1}$ into $[W]$. The mapping $\hat{F}_o = \hat{A}^{-1}\hat{F}$ is again a collineation on $\mathbb{P}^n$ and in addition, $\hat{F}_o$ leaves invariant the hyperplane at infinity $\mathbb{P}^{n-1}$: The map $\hat{F}$ transforms $\mathbb{P}^{n-1}$ to $[W]$, and $\hat{A}^{-1}$ maps $[W]$ back to $\mathbb{P}^{n-1}$. Thus $\hat{F}_o$ falls under the special case discussed at the beginning and is therefore semiprojective. Thus also $\hat{F} = \hat{A}\hat{F}_o$ is semiprojective, being a composition of semiprojective mappings. □

Remark

1. For $\mathbb{K} = \mathbb{R}$, semi-projective and projective mappings are the same, because $\mathbb{R}$ does not admit (continuous) automorphisms.
2. Using a similar argument, we can also prove the following local version:

   If $U_1, U_2 \subset \mathbb{R}^n \subset \mathbb{RP}^n$ are open sets and $F : U_1 \to U_2$ is a straight and invertibly continuous mapping, then F is a restriction of a projective mapping.

   For this we consider a hyperplane $H_1 \subset \mathbb{R}^n$ which intersects U_1. The image $F(H_1 \cap U_1)$ lies in another hyperplane H_2 intersecting U_2, since F maps lines to lines. We then choose two projective mappings F_1, F_2 on $\mathbb{P}^n$ which map H_1, H_2 onto the hyperplane at infinity $\mathbb{P}^{n-1}$. The composition $\tilde{F} = F_2^{-1}FF_1$ defined on the open subset[14] $\tilde{U}_1 = F_1^{-1}(U_1) \subset \mathbb{P}^n$ is then parallel on $\tilde{U}_1 \cap \mathbb{A}^n$ and thus a restriction of an affine mapping, see the remark after Theorem 2.1. So $F = F_2\tilde{F}F_1^{-1}$ is the restriction of a projective mapping (a composition of projective mappings).
3. A corresponding theorem with analogous proof also holds for straight (i.e. mapping lines to lines or points) but no longer bijective mappings F between projective spaces of different dimension. However, non-injective semiprojective mappings $F = [F_o]$ are no longer defined on all of $\mathbb{P}^n$: The kernel of the associated semilinear map F_o is mapped onto the origin, and $[F_o][v] = [F_o v]$ is not defined when $F_o v = 0$.

[14] The real projective space $\mathbb{RP}^n$ has a *metric*, a notion of distance: Elements of $\mathbb{RP}^n$ are straight lines through the origin 0, and the distance between two such lines will be the angle between them. A subset $U \subset \mathbb{RP}^n$ is *open* if with every $[v] \in U$, the ball $B_\epsilon([v]) = \{[w];\ \angle(v, w) < \epsilon\}$ lies entirely in U.

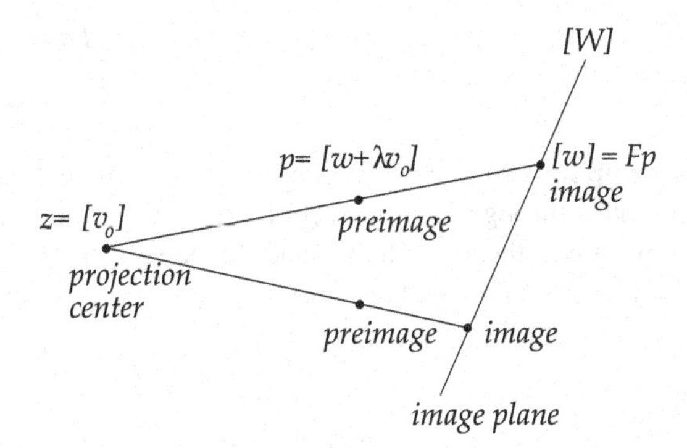

An illustrative example is the *central projection* onto a hyperplane $[W] \subset \mathbb{P}^n$ through a center $z = [v_o] \in \mathbb{P}^n \setminus [W]$ where the image of a point $p \in \mathbb{P}^n \setminus \{[v_o]\}$ is the intersection of the straight line pz with the hyperplane $[W]$. The corresponding semilinear mapping F_o is actually linear: It is the projection onto the W-component with respect to the direct decomposition $\mathbb{K}^{n+1} = W \oplus \mathbb{K}v_o$, that is $F_o(w + \lambda v_o) = w$; in fact, the image of $[F_o]$ is contained in $[W]$, and the three points $[w + \lambda v_o]$, $[w]$, $[v_o]$ (pre-image, image, center of projection) lie on a common straight line, as it should be. The kernel of F_o is the one-dimensional linear subspace $\mathbb{K}v_o$, and in fact we know that the central projection is not definable at the center $[v_o]$.

3.5 Theorem of Desargues

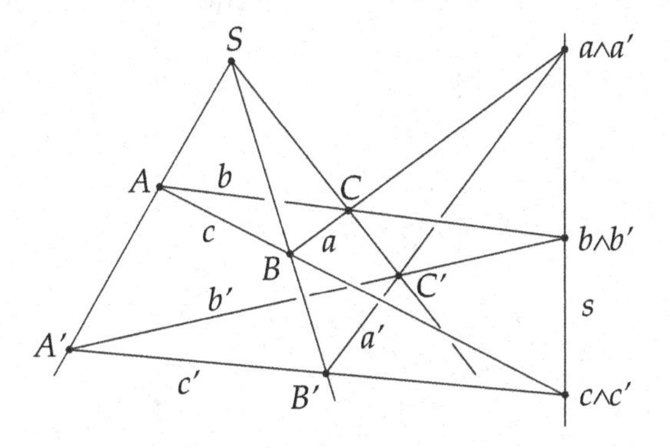

Theorem 3.3 (Desargues) *In the projective plane $\mathbb{P}^2$ (over any field $\mathbb{K}$, even over a skew field) we consider two triangles ABC and $A'B'C'$ with the property that the straight lines AA', BB', and CC' pass through a common point S. Then the three intersection points of corresponding sides of the two triangles (the three pairs of*

lines $c = AB$, $c' = A'B'$ *as well as* $b = AC$, $b' = A'C'$, *and* $a = BC$, $a' = B'C'$) *lie on a common straight line* s.

Proof 1 (With Mapping Geometry) Using a projective mapping, we can transform the straight line s through the intersection points $c \wedge c'$ and $b \wedge b'$ to the line at infinity. Then we are in the affine plane, and the pairs of lines c, c' and b, b' are parallel, since they meet on the line at infinity.

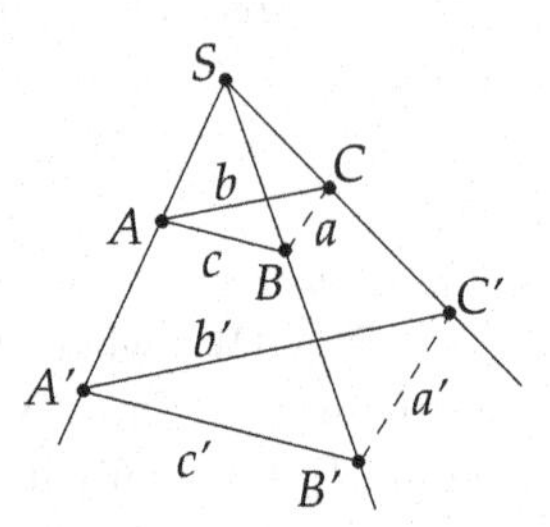

Thus the triangle $A'B'C'$ arises from the triangle ABC by a homothety with center S. Therefore also the third pair of lines a, a' is parallel, i.e. the lines a and a' intersect on the line at infinity. By back-transforming the line at infinity to s again, the assertion follows. □

Proof 2 (With Spatial Geometry) We can view the Desargues figure as a projection of a spatial figure, imagining, for example, that the middle straight line BB' is further ahead than AA' and CC'. The two triangles ABC and $A'B'C'$ now define two different planes E and E' in space $\mathbb{P}^3$ which always meet in a straight line s.[15]

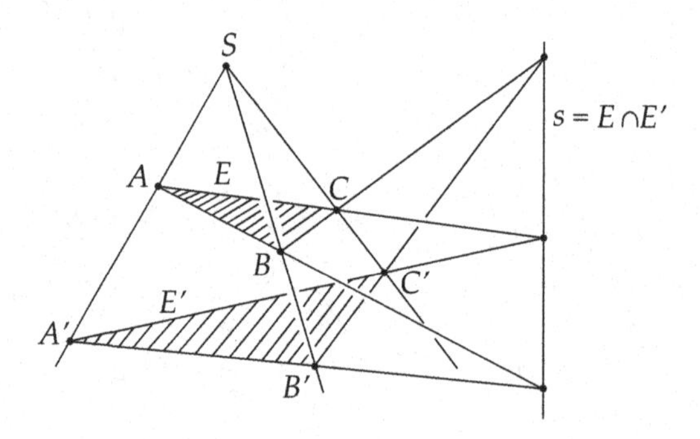

[15] This is the dimension theorem for linear subspaces $U, U' \subset V$: When $V = U + U'$ then $\dim(U \cap U') = \dim U + \dim U' - \dim V$. In our case $E = [U]$ and $E' = [U']$ we obtain $\dim U = \dim U' = 3$ and therefore $\dim U \cap U' = 3+3-4 = 2$, therefore $[U \cap U']$ is a projective straight line.

The prolongated sides of the two triangles lie in the planes E and E', respectively, so their intersection points (if existent) lie on $s = E \cap E'$. However, two straight lines in (projective) space intersect only if they lie in a common plane; but this is the case for corresponding sides of the two triangles: E.g., $c = AB$ and $c' = A'B'$ lie in the plane through S spanned by the rays AA' and BB'. □

Remark What is interesting about Proof 2 is that it only uses incidence, however in dimension 3. If one describes planar projective geometry by axioms,[16] then Desargues' theorem cannot be deduced from these axioms, but in spatial geometry it does follow from the axioms[17]; this is another example of our observation made earlier (Footnote 5 in Chap. 2) that geometry becomes simpler in higher dimensions. Thus, if a projective plane (defined axiomatically) can be extended to a projective space, it is a *Desargues plane,* i.e. Desargues' theorem holds in addition to the axioms in Footnote 16. This gives us the *homotheties*, i.e. the multiplication by scalars (cf. Sect. 2.1) is available, and thus we have geometrically reconstructed the scalar field of linear algebra. Therefore, the following theorem holds:

Every Desargues plane is of the form $\mathbb{KP}^2$ for a field or a skew field $\mathbb{K}$. Any (axiomatically defined) projective space is of the form $\mathbb{KP}^n$, $n \geq 3$.

There are projective planes *without* the Desargues property.[18] The most interesting example is the projective plane $\mathbb{OP}^2$ over the *octonions* $\mathbb{O}$: Similar to the complex numbers $\mathbb{C}$ as pairs of real numbers $(a, b) = a + bi$ with the multiplication $(a, b)(c, d) = (ac - bd, ad + bc)$ one can define the *quaternions* $\mathbb{H}$ as pairs of complex numbers with the multiplication $(a, b)(c, d) = (ac - b\bar{d}, ad + b\bar{c})$ (see Exercise 34) and the *octonions* $\mathbb{O}$ as pairs of quaternions with the multiplication $(a, b)(c, d) = (ac - \bar{d}b, da + b\bar{c})$. In each of these three processes, one must give up familiar rules of computation: In $\mathbb{C}$ there is no more ordering (< and >), in $\mathbb{H}$, multiplication is no longer commutative, and in $\mathbb{O}$ it is not even associative. Once more the process cannot be done without destroying division; therefore the octonions form the irrevocable last extension of number.[19] Because of the lack

---

16

(1) Two distinct points lie on exactly one line.
(2) Two distinct lines intersect at exactly one point.
(3) Every line contains ≥ 3 points and through each point pass ≥ 3 lines.

17 In space, axiom (2) must be supplemented by the addition of "two straight lines in a common plane" in order to exclude pairs of skew straight lines that are not in the same plane; such straight lines do not intersect, after all.

18 Cf. Salzmann et al: *Compact Projective Planes,* de Gruyter 1995.

19 Cf. Ebbinghaus et al. [6] or Eschenburg [12], Chap. 10.

of associativity, projective or affine *space* of dimension ≥ 3 cannot be defined over the octonions; the usual linear algebra no longer applies over the octonions. But what remains of it just suffices to define a projective *plane* over $\mathbb{O}$. In this, Desargues' theorem does not hold; it cannot be extended to a projective space. The rudimentary linear algebra over $\mathbb{O}$ is also responsible for the existence of the so-called *exceptional* groups G_2, F_4, E_6, E_7, E_8 which, unlike the *classical* groups $GL(n)$, $O(n)$, $U(n)$, $Sp(n)$ cannot be inserted into an infinite succession; the (non-compact version of the) group E_6, for example, is the collineation group of $\mathbb{OP}^2$. Some physicists are convinced that E_8, the largest and most mysterious of these groups, is deeply responsible for the structure of our material world.[20]

3.6 Conic Sections and Quadrics; Homogenization

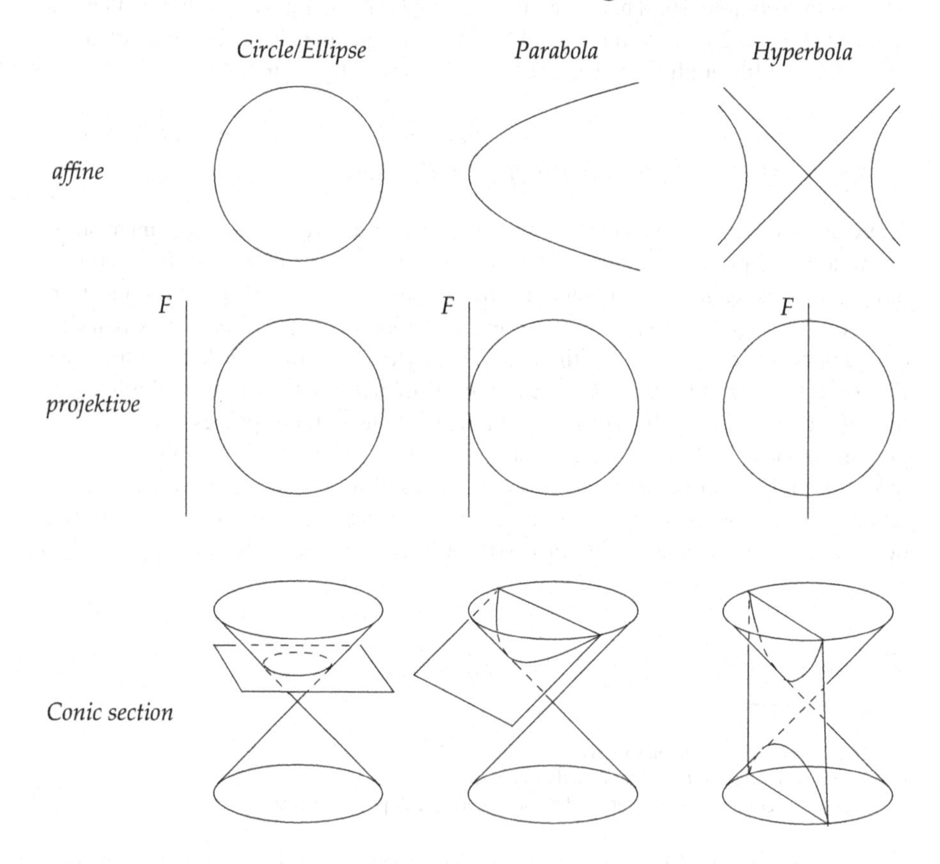

[20] See, for example https://aimath.org/E8/e8andphysics.html.

In three-dimensional space, if you intersect the cone over a circle with a plane,[21] depending on the position of the plane we get three kinds of intersection curves (*"conic section"*): ellipses (with the circle as a special case), parabolas and hyperbolas, and in addition, when the position of the plane is very special, also points, straight lines and straight line pairs.[22] If we leave these latter cases out of consideration, we are left with ellipses, parabolas and hyperbolas. These three kinds of conic sections are affinely different: We cannot transform an ellipse into a parabola or hyperbola by any *affine* mapping of the plane, if only because the ellipse is closed and the hyperbola splits into two parts ("branches"). But there are *projective* mappings that do this, once we extend affine plane to projective plane. In projective plane the three conic sections are kindred figures, namely simply closed curves, however in the three cases the line at infinity (F) lies at different positions with regard to the conic section: The ellipse does not intersect it, the parabola touches it, and the hyperbola intersects it at two points. So the parabola can be completed to a closed curve by just *one* point at infinity (the direction of the axis) while the hyperbola needs *two* points at infinity (the directions of the two asymptotes).

Strictly speaking, the term "conic section" already says that. If we put the apex of the cone at the origin 0 in linear 3-space then the cone is the union of a set of straight lines through 0 (the *generating lines* of the cone). But straight lines through 0 in space are points in the projective plane; hence the cone can be seen as a family of points in $\mathbb{P}^2$, hence as a *curve* $C \subset \mathbb{P}^2$. This is the common object in all three cases, "the" conic section. Ellipse, parabola and hyperbola are just the affine parts of C for different choices of the line at infinity in $\mathbb{P}^2$.

How can we explain these geometric observations *analytically* (with the help of formulas) and thus generalize them to arbitrary dimensions and for arbitrary fields $\mathbb{K}$? Analytically speaking, an (affine) conic section C_a is the solution set of a quadratic equation in two variables x and y, i.e. $C_a = \{(x, y);\ ax^2 + bxy + cy^2 + dx + ey + f = 0\}$. If we go from two variables x and y to n variables $x_1, \dots, x_n$ which we combine to one vector-valued variable $x = (x_1, \dots, x_n)$, the general quadratic equation is: $q(x) = 0$, where q stands for any quadratic expression in the coordinates $x_1, \dots, x_n$,

$$q(x) = \sum_{i,j=1}^{n} a_{ij} x_i x_j + \sum_{i=1}^{n} b_i x_i + c. \tag{3.9}$$

The solution set of a quadratic equation in n variables,

$$Q_{a'} = \{x \in \mathbb{K}^n;\ q(x) = 0\} \tag{3.10}$$

[21] For example, you can make the cone of light from a lamp fall on a wall.

[22] Another special case is a pair of parallel straight lines. However, this does not occur by intersecting a plane with a cone, but only with a cylinder.

is called an affine *quadric*. In linear algebra, you learn that you can solve the quadratic equation $q(x) = 0$ by affine substitutions $x = A\tilde{x} + a$ and reduce it to a few standard equations (*normal forms*). For $n = 2$ and $\mathbb{K} = \mathbb{R}$ the three most important ("non-degenerate") cases are the equations of the circle, $x^2 + y^2 - 1 = 0$, the hyperbola, $x^2 - y^2 - 1 = 0$ and the parabola, $x^2 - y = 0$.

How can we extend a quadric from affine to projective space? We need to look at the polynomial q in (3.9) a little closer. We decompose q into its three parts $q = q_2 + q_1 + q_0$: the quadratic part $q_2(x) = \sum_{ij} a_{ij}x_ix_j$, the linear part $q_1(x) = \sum_i b_ix_i$ and the constant $q_0 = c$. The three parts obviously behave differently if we substitute x by λx for a scalar $\lambda \in \mathbb{K}$: Then $q_k(\lambda x) = \lambda^k q_k(x)$ for $k = 2, 1, 0$. In general, a function $f : \mathbb{K}^n \to \mathbb{K}$ with $f(\lambda x) = \lambda^k f(x)$ for all $\lambda \in \mathbb{K}$ is called *homogeneous of degree* k, the three parts q_0, q_1, q_2 of our quadratic polynomial q are thus homogeneous of degree 0, 1, 2. Every polynomial is a sum of homogeneous polynomials.

We consider $\mathbb{K}^n$ now as the affine part $\mathbb{A}^n = \{[x, 1];\ x \in \mathbb{K}^n\}$ of the projective space $\mathbb{P}^n$ and set $Q_a = \{[x, 1];\ q(x) = 0\}$. For a point $[x, \xi]$ with $\xi \neq 0$, that is $[x, \xi] = [\frac{x}{\xi}, 1] \in \mathbb{A}^n \subset \mathbb{P}^n$, we have

$$\begin{aligned}
[x, \xi] \in Q_a \iff 0 &= q(\frac{x}{\xi}) = q_2(\frac{x}{\xi}) + q_1(\frac{x}{\xi}) + q_0(\frac{x}{\xi}) \\
&= \frac{1}{\xi^2} q_2(x) + \frac{1}{\xi} q_1(x) + q_0(x) \\
\iff 0 &= q_2(x) + \xi q_1(x) + \xi^2 q_0(x).
\end{aligned}$$

The right-hand side of the last equation, the expression

$$\hat{q}(x, \xi) := q_2(x) + \xi q_1(x) + \xi^2 q_0(x),$$

is a homogeneous polynomial of degree 2 in the $n + 1$ variables $x_1, \ldots, x_n, \xi$, and the expression $\hat{q}(x, \xi)$ makes sense also for $\xi = 0$ and thus for all $[x, \xi] \in \mathbb{P}^n$. Thus we have found the projective extension, also called *projective closure* Q of Q_a:

$$Q = \{[\hat{x}] \in \mathbb{P}^n;\ \hat{q}(\hat{x}) = 0\}. \tag{3.11}$$

This transition from q to $\hat{q}$ is called *homogenization*: A polynomial f of degree d in n variables becomes a homogeneous polynomial $\hat{f}$ of the same degree d in $n + 1$ variables. One decomposes f first into its homogeneous components, $f = \sum_{k=0}^{d} f_k$, where f_k is homogeneous of degree k. Then one multiplies f_k by the $(d - k)$-th power of a new variable ξ or x_{n+1} and obtains a polynomial $\hat{f}$ in $n + 1$ variables which can be combined into one $\mathbb{K}^{n+1}$-valued variable $\hat{x} = (x_1, \ldots, x_{n+1})$:

$$\hat{f}(\hat{x}) = \sum_{k=0}^{d} (x_{n+1})^{d-k} f_k(x_1, \ldots, x_n).$$

In fact, $\hat{f}$ is homogeneous of degree d, because

$$\hat{f}(\lambda\hat{x}) = \sum_k (\lambda x_{n+1})^{d-k} f_k(\lambda x) = \lambda^d f(\hat{x}).$$

Example with $n = d = 2$: For $f(x, y) = x^2 + 2xy - y^2 + 2x - 1$ we have

$$\hat{f}(x, y, z) = x^2 + 2xy - y^2 + 2xz - z^2.$$

Now we can extend the zero set $N_a = \{x \in \mathbb{K}^n;\ f(x) = 0\}$ from affine to projective space by putting

$$N = \{[\hat{x}] \in \mathbb{P}^n;\ \hat{f}(\hat{x}) = 0\}.$$

Because of the homogeneity of $\hat{f}$ this set is "well-defined": The validity of the equation $\hat{f}(\hat{x}) = 0$ does not depend on which representative of the equivalence class $[\hat{x}]$ we choose, because for every scalar $\lambda \neq 0$ we have

$$\hat{f}(\lambda\hat{x}) = \lambda^d \hat{f}(\hat{x}) = 0 \iff \hat{f}(\hat{x}) = 0.$$

Moreover, $N \cap \mathbb{A}^n = N_a$, since $\hat{f}(x, 1) = f(x)$ for all $x \in \mathbb{K}^n$.

In Exercise 19 we see directly that the projective closures of the ellipse, parabola, and hyperbola are projectively equivalent. Here we show generalization for arbitrary dimension:

Theorem 3.4 *Any quadric $Q \subset \mathbb{P}^n$ is projectively equivalent to the solution set of one of the equations*

$$\sum_{i=1}^{m} \epsilon_i x_i^2 = 0 \tag{3.12}$$

for numbers $\epsilon_1, \ldots, \epsilon_m \in \mathbb{K}^*$ and $0 \leq m \leq n + 1$. For $\mathbb{K} = \mathbb{R}$ ($\mathbb{K} = \mathbb{C}$) one can choose $\epsilon_i = \pm 1$ ($\epsilon_i = 1$) for all $i = 1, \ldots, m$.

Proof Let $V = \mathbb{K}^{n+1}$ and $Q = \{[x] \in \mathbb{P}^n;\ q(x) = 0\}$ for some homogeneous quadratic polynomial (*quadratic form*) q (we now omit the notations $\hat{q}$ and $\hat{x}$). We need to show that there is an invertible linear mapping A on V with $q(A(x)) = \sum_i \epsilon_i x_i^2$. This is familiar from linear algebra: q belongs to a symmetric bilinear form $\beta : V \times V \to \mathbb{K}$ with $q(x) = \beta(x, x)$, which is obtained by *polarization*[23]:

$$\beta(x+y, x+y) = \beta(x,x) + \beta(y,y) + 2\beta(x,y) \Rightarrow$$

$$2\beta(x,y) = q(x+y) - q(x) - q(y).$$

We show by induction over n that there is a basis $b_1, \dots, b_{n+1}$ with $\beta(b_i, b_j) = 0$ for $i \neq j$. To do this, we just look for a vector b with $q(b) \neq 0$ (which exists because $q \neq 0$) and set $V' = \{x \in V;\ \beta(x,b) = 0\}$. This linear subspace has one dimension less (n instead of $n+1$), so according to the induction assumption there is a basis $b_1, \dots, b_n$ of V' with $\beta(b_i, b_j) = 0$ for $i \neq j$. The sought basis of V we obtain by adding $b_{n+1} := b$. (The beginning of induction at $n = 0$ is trivial). If now one chooses A as the linear mapping with $A(e_i) = b_i$, i.e. $A = (b_1, \dots, b_{n+1})$ as matrix, then $q(Ax) = q(A(\sum_i x_i e_i)) = q(\sum_i x_i b_i) = \beta(\sum_i x_i b_i, \sum_j x_j b_j) = \sum_{ij} x_i x_j \beta(b_i, b_j) = \sum_i \epsilon_i x_i^2$ with $\epsilon_i = \beta(b_i, b_i) = q(b_i)$. If you now omit all the summands for which $\epsilon_i = 0$ and renumber the coordinates accordingly, you get the normal form (3.12). When $\mathbb{K} = \mathbb{C}$, you can take a square root of ϵ_i for $i = 1, \dots, m$ and renormalize the elements of the basis to $\tilde{b}_i = b_i/\sqrt{\epsilon_i}$, so $\tilde{\epsilon}_i := q(\tilde{b}_i) = 1$. For $\mathbb{K} = \mathbb{R}$ you can at least take the square root of $|\epsilon_i|$, and for $\tilde{b}_i = b_i/\sqrt{|\epsilon_i|}$ holds $\tilde{\epsilon}_i = q(\tilde{b}_i) = \pm 1$. □

Corollary 3.5 *In $\mathbb{RP}^n$ there are precisely $[\frac{n+1}{2}]$ non-degenerate quadrics ($m = n+1$), up to projective equivalence, and their equations are*

$$x_1^2 + \cdots + x_p^2 - x_{p+1}^2 - \cdots - x_{n+1}^2 = 0$$

with $1 \leq p \leq [\frac{n+1}{2}]$. In particular, for $n = 2$ there is just one, the conic section with the equation $x^2 + y^2 - z^2 = 0$, and for $n = 3$ there are two, called Q_1, Q_2, with the equations $x^2 + y^2 + z^2 - w^2 = 0$ and $x^2 + y^2 - z^2 - w^2 = 0$.

Proof The normal form of the quadratic equation is $\pm(x_1)^2 \pm \ldots \pm (x_{n+1})^2 = 0$. If we rearrange the coordinates so that the negative terms come last, there are $n+2$ possibilities (0 negative terms to $n+1$ negative terms). But since we can multiply the whole equation by -1 we may assume that there are no more positive terms than negative ones.[24] Moreover, the equation with exclusively positive (or exclusively negative) terms has only the zero solution, to which no point in $\mathbb{P}^n$ corresponds to; the solution set of this equation in $\mathbb{P}^n$ is therefore empty. The cases given remain. □

[23] Here we must assume once again $\mathrm{char}(\mathbb{K}) \neq 2$ since otherwise $2\beta(x,y) = 0$.

[24] This is the easiest convention for counting the quadrics. However, the usual convention is that there are not more negative than positive terms.

In Sect. 3.9 we will present a constructive procedure for determining quadrics in $\mathbb{P}^n$.

We still want to look a little closer at the two quadrics Q_1 and Q_2 in $\mathbb{RP}^3$. The quadric Q_1 with the equation $x^2+y^2+z^2 = w^2$ is a sphere: One can assume $w \neq 0$ and hence $w = 1$ because $w = 0$ would imply $x = y = z = 0$; thus the quadric does not intersect the plane at infinity $\{w = 0\}$ and therefore it lies entirely in the affine part $\mathbb{A}^3$. The other quadric Q_2 with the equation $x^2 + y^2 - z^2 = w^2$ has as affine part the one-sheeted hyperboloid $x^2 + y^2 - z^2 = 1$ and intersects the plane at infinity $\mathbb{P}^2 = \{w = 0\}$ in the circle $x^2 + y^2 = z^2$ (in $\mathbb{P}^2$ one has to put z to one in order to see the circle equation, while $z = 0$ does not give a solution). All other non-degenerate affine quadrics are projectively equivalent to one of these. For example, the two-sheeted hyperboloid $x^2 - y^2 - z^2 = 1$ is equivalent to Q_1: Homogenizing the equation yields $x^2 - y^2 - z^2 - w^2 = 0$, thus $y^2 + z^2 + w^2 = x^2$; this is the equation of Q_1 with reversed roles of w and x. The plane at infinity $\{w = 0\}$ intersects this quadric in the "circle" $x^2 = y^2 + z^2$ and divides it into two parts, the two sheets of the two-sheeted hyperboloid.

Let us investigate the quadric Q_2, where the field $\mathbb{K}$ may now be arbitrary. We have seen that the affine part of Q_2 is the one-sheeted hyperboloid, on which, as is well known, there are two families of straight lines.

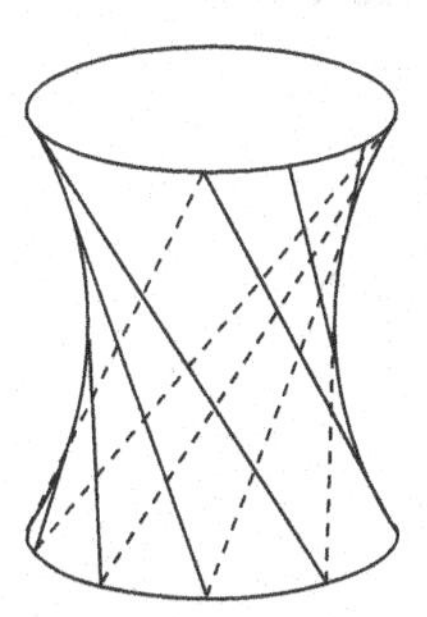

We can see this particularly easily in the projective model: The equation of Q_2 is $x^2 - z^2 = w^2 - y^2$, that is $(x + z)(x - z) = (w + y)(w - y)$. The four expressions $x \pm z$, $w \pm y$ can be used as new coordinates s, t, u, v; the coordinate transformation is an invertible linear map, so it gives a projective mapping. The equation of Q_2 thus becomes

$$st = uv.$$

Special solutions are $s/u = v/t = \alpha$ and likewise $s/v = u/t = \beta$ for constants $\alpha, \beta \in \hat{\mathbb{K}} = \mathbb{K} \cup \{\infty\}$ (where $s/u = \infty$ just means $u = 0$). These are two linear equations that describe a two-dimensional linear subspace of $\mathbb{K}^4$ and thus a straight line in $\mathbb{P}^3$. The numbers α and β thus parametrize two families of straight lines, which lie entirely on Q_2 because all their points satisfy the equation $st = uv$ for Q_2.

Remark We can identify $\hat{\mathbb{K}}$ with $\mathbb{P}^1$ using the mapping

$$\mathbb{P}^1 \to \hat{\mathbb{K}}, \ [\alpha_1, \alpha_2] \mapsto \alpha := \alpha_1/\alpha_2. \tag{3.13}$$

Thus the quadric $Q_2 \subset \mathbb{P}^3$ becomes $\mathbb{P}^1 \times \mathbb{P}^1$ using the bijective mapping $s : \mathbb{P}^1 \times \mathbb{P}^1 \to Q_2$,

$$s([\alpha_1, \alpha_2], [\beta_1, \beta_2]) = [s, t, u, v] := [\alpha_1\beta_1, \alpha_2\beta_2, \alpha_2\beta_1, \alpha_1\beta_2], \tag{3.14}$$

the *Segre embedding*.[25] The image of s actually lies in Q_2, for the equation $st = uv$ is satisfied because $\alpha_1\beta_1\alpha_2\beta_2 = \alpha_2\beta_1\alpha_1\beta_2$ (using commutativity of the multiplication in $\mathbb{K}$). Obviously, we obtain a straight line if we set the first argument $[\alpha_1, \alpha_2]$ constant. This it is the first of the above two families of straight lines, $\alpha = const$, because $s/u = v/t = \alpha_1/\alpha_2$. Setting the second argument $[\beta_1, \beta_2]$ constant we get the second family of straight lines $\beta = const$.

3.7 Theorem of Brianchon

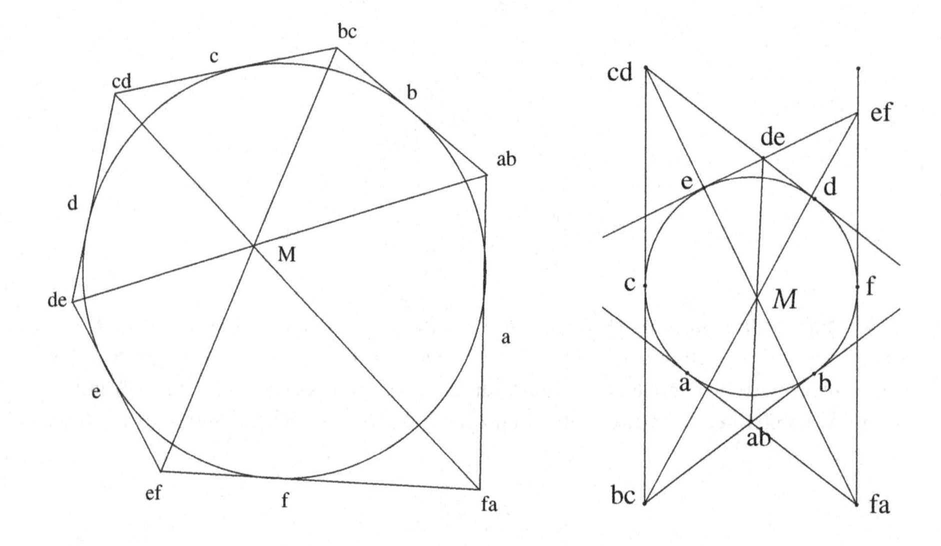

Theorem 3.6 of Brianchon[26]: *In the projective plane $\mathbb{P}^2$ the three diagonals of a hexagon, whose sides are tangents*[27] *of a conic section, intersect at a common point.*

[25] Beniamino Segre, 1903 (Turin)–1977 (Frascati).

[26] Charles Julien Brianchon, 1783 (Sèvres near Paris)–1864 (Versailles).

[27] For the precise definition of a tangent see below.

Proof The proof is similar to the second proof of Desargues' theorem: It represents the tangent hexagon of the conic section as a projection of a spatial figure. This figure consists of certain straight lines on a one-sheeted hyperboloid. To simplify matters we first transform the conic section by a projective mapping onto the circle $K = \{(x, y);\ x^2 + y^2 = 1\}$ in the xy-plane E contained in affine (linear) xyz-space. In this space we have the one-sheeeted hyperboloid $H = \{(x, y, z);\ x^2 + y^2 - z^2 = 1\}$. The *tangent plane* of H at each point $P = (x, y, 0) \in K = H \cap E$ is vertical (parallel to z-axis); it contains the tangent at P of any curve on H through P (see below), in particular the tangent to the circle K at P and the two straight lines $t \mapsto (x + ty, y - tx, \pm t)$ through P, which run completely on H (cf. Sect. 3.6 and Exercise 23). The orthogonal projection onto E maps both (*ascending* and *descending*) straight lines onto the tangent of K in P. Now we replace the planar hexagon by a spatial hexagon which consists of alternately ascending and descending line segments on H and which is projected onto the given hexagon (see left figure below).

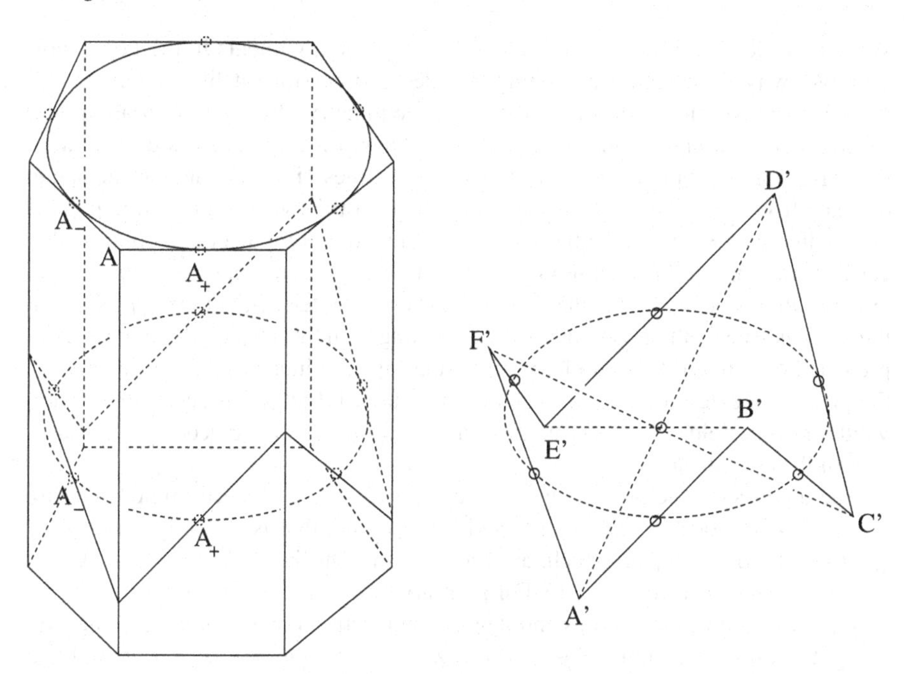

Since each vertex A of the tangent hexagon is equidistant to the points of contact $A_+, A_- \in K$ of the two tangents, the ascending straight line through A_+ and the descending straight line through A_- meet at the same height z in a common point A' above the vertex A (right figure). We call the vertices of the tangent hexagon $A, \ldots, F$, and the vertices of the spatial hexagon above or below these points are called $A', \ldots, F'$. We have to show that the diagonals of the spatial hexagon, $A'D'$ and $B'E'$ and $C'F'$, meet at a common point; then the same is true for the diagonals of the planar hexagon. But with the spatial hexagon we have introduced a new problem that did not yet exist with the planar hexagon: In space, after all, it is not even clear that at least two of the three diagonals meet! But the solution of this new problem will solve the actual task of finding a common intersection of all three diagonals. So why do the diagonals $A'D'$ and $B'E'$ meet? Because they lie in a common plane! Since any ascending and descending lines intersect (see below), possibly in a point at infinity, also $A'B'$ and $D'E'$ have a point of intersection, because one of these lines is ascending, the other one is descending. So the four points A', B', D', E' are *coplanar* (contained in a common plane), and so the lines in question $A'D'$ and $B'E'$ lie in this plane and must therefore intersect. Likewise, the pairs of straight lines $B'E'$ and $C'F'$ as well as $C'F'$ and $D'A'$ intersect. If the three intersection points are distinct, they form a planar triangle, and all points and straight lines lie in a common plane. But this cannot be, because the spatial hexagon does not lie in a common plane (which would necessarily be the plane E of the circle K). So all three diagonals must pass through the same point, and the same must happen for the projections onto E. □

We need to look a little more closely at some of the arguments in the proof. For example, why do arbitrary ascending and descending straight lines intersect? In the figure above, this follows for reasons of symmetry: The reflection along the vertical plane through A and the z-axis interchanges A_+ with A_- and also swap the corresponding tangents at K in the xy-plane; these intersect at A at the same distance from $A_\pm$. The straight lines through A on the hyperboloid, ascending or descending with slope 1 over these tangents, must therefore meet at a point on the vertical straight line through A. Only when A_+ and A_- are opposite points of the circle K, that is $A_- = -A_+$, the two tangents do not intersect, but are parallel, and the same is true of the ascending and descending straight lines on the hyperboloid projected onto them. But parallel straight lines also lie in a common plane, and in the projective extension they intersect (at a point at infinity). These arguments are valid for every field, where the square of the distance is to be replaced by the square sum of the coordinates.

The algebraic structure becomes more apparent if we proceed immediately to the projective model of the one-sheeted hyperboloid, that is to the quadric $Q = \{[s,t,u,v];\ st = uv\}$ on which, after all, the straight lines $s/u = v/t = \lambda$ and $s/v = u/t = \mu$ lie (see Sect. 3.6). For arbitrary $\lambda, \mu \in \mathbb{K} \cup \{\infty\}$ we find a solution (s, t, u, v) for all four equations and thus an intersection point $[s, t, u, v]$ of the two straight lines. For example, if $\mu \neq 0$ and $\lambda \neq \infty$, then $u = 1$, $s = \lambda$, $t = 1/\mu$, $v = \lambda/\mu$.

Another question is conceptual: What actually is the tangent of a conic section and the tangent plane of a quadric? The definition is simple: Every projective quadric in $\mathbb{P}^n$ is given by a quadratic form q on $\mathbb{K}^{n+1}$ and thus by a symmetric bilinear form β with $q(x) = \beta(x,x)$ (cf. Sect. 3.6). Thus, if

$$Q = \{[x] \in \mathbb{P}^n;\ \beta(x,x) = 0\}, \tag{3.15}$$

so the *tangent space* of Q at a point $[x] \in Q$ is the hyperplane

$$T_{[x]}Q := \{[v] \in \mathbb{P}^n;\ \beta(x, v) = 0\}. \tag{3.16}$$

In the plane ($n = 2$) the tangent space is called *tangent*, in 3-space ($n = 3$) we call it *tangent plane*. In the example of the one-sheeted hyperboloid $Q = \{[x, y, z, w];\ x^2 + y^2 - z^2 - w^2 = 0\}$ we have

$$\beta((x, y, z, w), (x', y', z', w')) = xx' + yy' - zz' - ww',$$

and the tangent plane at a point $p = [x, y, 0, 1] \in K \subset Q$ is

$$T_pQ = \{[x', y', z', w'];\ xx' + yy' - w' = 0\}$$

with the affine part $T_pQ \cap \mathbb{A}^n = \{[x', y', z', 1]; xx' + yy' = 1\}$; since z' does not occur in the equation, it can be chosen arbitrarily, so this plane is parallel to the z-axis, as asserted.

You cannot prove definitions, but you can justify them. Why is the hyperplane (3.16) the *tangential* (i.e. touching) hyperplane or, depending on the dimension, tangent (line), tangent plane, tangent space? Descriptively, it should approximate the quadric Q near its touching point $[x] \in Q$. Therefore, if we choose a second point $[x+v] \in Q$, then both $\beta(x, x) = 0$ as well as $\beta(x+v, x+v) = 0$, so

$$0 = \beta(x + v, x + v) = \beta(x, x) + 2\beta(x, v) + \beta(v, v) = 2\beta(x, v) + \beta(v, v).$$

If $[x + v]$ is now very close to $[x]$, at least in the cases $\mathbb{K} = \mathbb{R}, \mathbb{C}$, the components of v have very small absolute values and therefore the quadratic term $\beta(v, v)$ in the variable v is much smaller in absolute value than the linear term $2\beta(x, v)$. If we simply omit the quadratic term (this can be done over an arbitrary field $\mathbb{K}$), we obtain the equation of the tangent hyperplane.

3.8 Duality and Polarity; Pascal's Theorem

An observation already made by Poncelet is the *principle of duality:* With every theorem on planar projective geometry also the *dual* theorem holds, in which the words "point" and "line" as well as "intersect" and "connect" are interchanged. Already the three axioms (Footnote 16) have this property (two points are connected by exactly one straight line—two straight lines intersect in exactly one point), so duality also holds for the set of theorems derived from them. The same is true for higher dimensions, where we have to replace lines by hyperplanes.

However, we did not define a projective space by axioms, but we derived it from a vector space: If V is a vector space over a field $\mathbb{K}$, the corresponding projective space P_V is the set of *homogeneous vectors* $[v] = \{\lambda v;\ \lambda \in \mathbb{K}^*\}$ for all $v \in V \setminus \{0\}$, and the canonical projection $\pi : V \setminus \{0\} \to P_V,\ v \mapsto [v]$ relates linear algebra to projective geometry. Therefore, we can use the duality principle from linear algebra: A projective hyperplane $H \subset P_V$ corresponds to a linear hyperplane $U \subset V$ (a linear subspace of codimension one): $H = \pi(U) = [U]$. This in turn can be described as the *kernel* of a linear form $\alpha \in \mathrm{Hom}(V, \mathbb{K}) = V^*$, that is $U = \ker\alpha$. We understand α as an element of another vector space: the *dual space* $V^* = \mathrm{Hom}(V, \mathbb{K})$. Every multiple $\lambda\alpha$ of $\alpha \in V^*$ with $\lambda \neq 0$ has the same kernel U, so to U and thus to H corresponds a homogeneous vector $[\alpha] \in P_{V^*}$. The hyperplanes in P_V can therefore be understood as points of another projective space, the *dual projective space* P_{V^*}.

More generally, we can assign to each $(k+1)$-dimensional linear subspace $U \subset V$ (where $\dim V = n+1$) the $(n-k)$-dimensional subspace

$$U^\perp := \{\alpha \in V^*;\ \alpha|_U = 0\} \subset V^*\,, \tag{3.17}$$

and for two subspaces $U_1, U_2 \subset V$ holds:

$$\begin{aligned} (U_1 \cap U_2)^\perp &= (U_1)^\perp + (U_2)^\perp\,, \\ (U_1 + U_2)^\perp &= (U_1)^\perp \cap (U_2)^\perp. \end{aligned} \tag{3.18}$$

Thus, a $(k+1)$-dimensional projective subspace $[U] \subset P_V$ is transformed by duality to the $(n-k)$-dimensional subspace $[U^\perp] \subset P_{V^*}$, and the operations "intersecting" $\wedge$ (corresponding to $U_1 \cap U_2$) and "connecting" $\vee$ (corresponding to $U_1 + U_2$) are interchanged with each other, according to (3.18). Since $(V^*)^* = V$ for any finite-dimensional vector space V, the process is invertible. In particular for $k = 0$ we see the duality between points and hyperplanes of the projective space.

Example 1 (Desargues Theorem)
$S = AA' \wedge BB' \wedge CC' \Rightarrow AB \wedge A'B',\ AC \wedge A'C',\ BC \wedge B'C'$ are collinear.

Dual Theorem
$s = aa' \vee bb' \vee cc' \Rightarrow ab \vee a'b', ac \vee a'c', bc \vee b'c'$ have common intersection.[28]

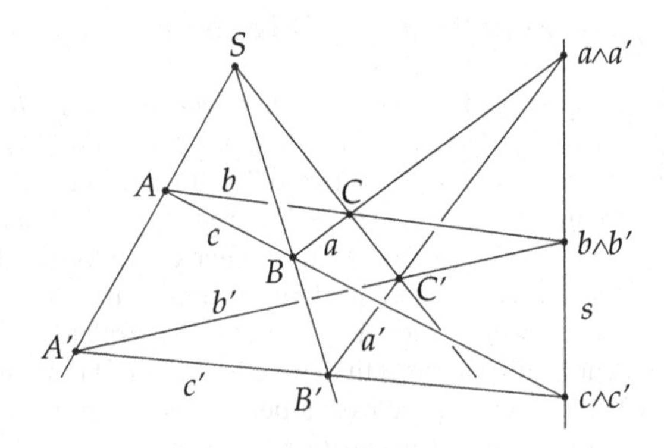

[28] Upper and lower case letters denote points and straight lines as before, $AB = A \vee B$ the straight line through the points A, B, and $ab = a \wedge b$ the intersection of the lines a, b.

The dual theorem in this case is just the inverse of Desargues' theorem. To see this, set $a = BC$, $b = CA$, $c = AB$ and correspondingly $a' = B'C'$, $b' = C'A'$, $c' = A'B'$ (see figure). Then Desargues' theorem says: *If the vertices of two triangle lie on straight lines passing through a common point S, then corresponding sides intersect on a common straight line s.* The dual theorem says: *If corresponding sides of two triangles intersect on a straight line s, then the vertices lie on straight lines through a common point S.*

Example 2 (Theorem of Brianchon)
If $a, \dots, f$ are tangents of a conic section, then the three straight lines $ab \vee de$ and $bc \vee ef$ and $cd \vee fa$ have a common intersection.

Dual Theorem: Theorem of Pascal[29]
If $A, \dots, F$ are points of a conic section, then the three points $AB \wedge DE$ and $BC \wedge EF$ and $CD \wedge FA$ are collinear.

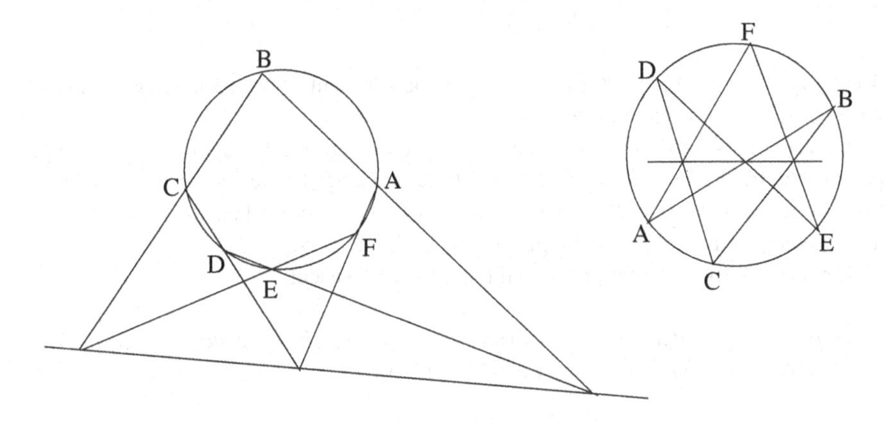

Lemma 3.7 *The tangents of a conic section in $\mathbb{P}^2$ are dual to the points of another conic section.*

Proof Let the given conic section be $Q = \{[x] \in \mathbb{P}^2;\ \beta(x, x) = 0\}$ for some non-degenerate symmetric bilinear form β on $V = \mathbb{K}^3$. For each $[x] \in Q$ we have $T_{[x]}Q = \{[v];\ \beta(x, v) = 0\} = \ker \beta_x$ where $\beta_x : v \mapsto \beta(x, v)$. These linear forms $\beta_x \in V^*$ with $[x] \in Q$ again satisfy a quadratic equation:

$$0 = \beta(x, x) = \beta_x(x) = \beta_x(\beta^{-1}(\beta_x)) = \beta^{-1}(\beta_x, \beta_x).$$

Here β is considered as an isomorphism $x \mapsto \beta_x : V \to V^*$ and its inverse $\beta^{-1} : V^* \to V$ can in turn be read as the bilinear form on V^* with $(\lambda, \mu) \mapsto \lambda(\beta^{-1}(\mu))$

[29] Blaise Pascal, 1623 (Clermont-Ferrand)–1662 (Paris).

for any $\lambda, \mu \in V^*$. So the corresponding homogeneous vectors $[\beta_x]$ belong to the *dual quadric*

$$Q^* := \{[\alpha];\ \alpha \in V^* \setminus \{0\};\ \beta^{-1}(\alpha, \alpha) = 0\} \subset P_{V^*}.$$

□

Any non-degenerate bilinear form $\beta : V \times V \to \mathbb{K}$ thus defines the vector space isomorphism $\beta : V \to V^*, x \mapsto \beta_x = (v \mapsto \beta(x, v))$. This defines an isomorphism of projective spaces $[\beta] : P_V \to P_{V^*}$. The two projective spaces P_V and P_{V^*} are thus identified by means of $[\beta]$ which makes the dual space dispensable: We can assign to each hyperplane $H = [\ker\alpha] \subset P_V$ the point $[\beta^{-1}(\alpha)] \in P_V$, and vice versa to each point $[x]$ the hyperplane $[x^\perp] = \{[y];\ \beta(x, y) = 0\} = [\ker\beta_x]$. These mappings are inverses of each other:

$$\begin{array}{ccccccc} [\ker\alpha] & \mapsto & [\beta^{-1}(\alpha)] & \mapsto & [\ker\beta_{\beta^{-1}(\alpha)}] & = & [\ker\alpha], \\ [x] & \mapsto & [\ker\beta_x] & \mapsto & [\beta^{-1}(\beta_x)] & = & [x], \end{array}$$

for $\beta_{\beta^{-1}(\alpha)} = \alpha$ and $\beta^{-1}(\beta_x) = x$. Such an assignment of points to hyperplanes in a projective space is called a *polarity.*

Instead of using duality, it is easier to prove Lemma 3.7 using *polarity*, e.g., polarity by the bilinear form β defining the quadric Q. In fact, polar to the straight line $T_{[x]}Q = \{y;\ \beta(x, y) = 0\} = [x]^\perp$ is then the point $[x]$; thus this polarity assigns the points of Q to the tangents of Q and vice versa.

Let us also give a direct proof of the theorem of Pascal:

Opposite edges of a hexagon inscribed in a conic section (e.g. circle or ellipse) intersect on a common straight line (left figure).

The proof reduces the assertion of the theorem to a simple special case using a projective transformation, similar to the first proof of Desargues' theorem:

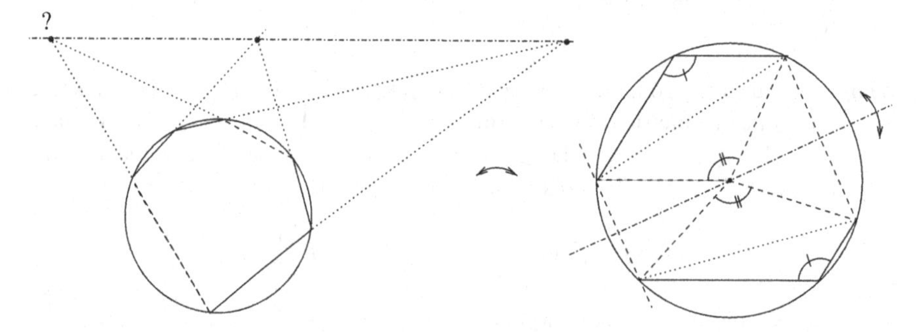

In fact, we can interpret the left figure as a perspective image of the right figure; the horizon is the horizontal straight line through the right two of the three intersection points of the edges. Thus, in the figure on the right, the two corresponding (undashed) edge pairs are parallel.[30] Hence the two (simply dashed) angles between the parallel pairs of straight lines in the right-hand figure are equal. Then, by the *inscribed angle theorem* (see below), the (doubly dashed) midpoint angles of the two dotted chords of the circle are equal. Thus, the two angles are mirror symmetric with mirror axis through the center of the circle (dash-dot straight line in the right figure). This reflection preserves the circle and maps the two pairs of radii onto each other, which span the two (doubly dashed) angles at the center. It thus interchanges the endpoints of each of the two dashed hexagon sides and keeps these invariant. Thus they must both be perpendicular to the mirror axis and thus parallel to each other. So all three pairs of hexagon-sides meet on the line at infinity and after back-transformation on the "horizon", which proves the assertion.

The *inscribed angle theorem* (figure below) states: *For each chord of the circle the central angle γ and the peripheral angle α (left figure) are in the relation* $\gamma = 360° - 2\alpha$ (for a variant see the solution of Exercise 30). This can be seen from the right figure: Since the base angles of a triangle formed by two radii and a chord are equal and the angle sum is 180°, we have $\gamma_1 + 2\alpha_1 = 180°$ and $\gamma_2 + 2\alpha_2 = 180°$, thus $\gamma + 2\alpha = 360°$.

In the spacial case when the chord is a diameter we have $\gamma = 180°$ and $\alpha = 90°$ (theorem of Thales).[31]

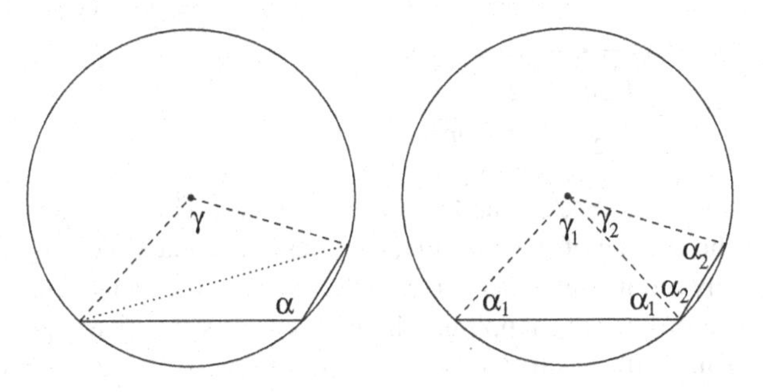

3.9 Projective Determination of Quadrics

How can we read from a quadratic equation, e.g. the equation

$$q(x, y, z, w) := x^2 - 2xy + 2z^2 + 4xz - 2yw = 0 \tag{3.19}$$

to which type of projective quadrics it belongs? There are two equivalent procedures which also can be considered as constructive proofs of Theorem 3.4:

[30] The circle may be distorted into an ellipse during the perspective transformation, but by compressing it in the direction of the longer ellipse axis, it becomes a circle again, and parallel pairs of lines remain parallel.

[31] Thales of Miletus, ca. 624–548 BC. The circle making a given line segment a diameter is called *Thales circle.*

1. *Quadratic completion:* We eliminate the mixed terms (e.g. the term $-2xy$) by variable substitutions after adding appropriate quadratic terms and subtracting them again. This creates squares of invertible linear expressions in the variables, and to these linear expressions we give new names, defining them as new variables. The old variables are now replaced by the inverse expressions in the new variables. The mixed terms are thereby replaced in lexicographic order (for the succession x, y, z, w) : xy, xz, xw, yz, yw, zw. For example, $x^2 - 2xy = x^2 - 2xy + y^2 - y^2 = (x-y)^2 - y^2$. Putting $x - y =: x_1$, it follows $x = x_1 + y$, and in our initial Eq. (3.19) we can replace (*substitute*) the variable x at each occurrence by $x_1 + y$. Here is the full procedure:

$$
\begin{aligned}
0 &= \underline{x^2 - 2xy} + 2z^2 + 4xz - 2yw \\
&\quad +y^2 - y^2 && x - y =: x_1, \quad x = x_1 + y \quad (a) \\
&= \underline{x_1^2} - y^2 + 2z^2 \underline{+4x_1 z} + 4yz - 2yw \\
&\quad \underline{+4z^2} - 4z^2 && x_1 + 2z =: x_2, \; x_1 = x_2 - 2z \; (b) \\
&= x_2^2 \underline{-y^2} - 2z^2 \underline{+4yz} - 2yw \\
&\quad \underline{-4z^2} + 4z^2 && y - 2z =: y_1, \;\; y = y_1 + 2z \;\; (c) \\
&= \underline{x_2^2 - y_1^2 - 2y_1 w} - 4zw + 2z^2 \\
&\quad \underline{-w^2} + w^2 && y_1 + w =: y_2, \;\; y_1 = y_2 - w \;\; (d) \\
&= x_2^2 - y_2^2 \underline{-4zw + 2z^2} + w^2 \\
&\quad \underline{+2w^2} - 2w^2 && z - w =: z_1, \quad z = z_1 + w \quad (e) \\
&= x_2^2 - y_2^2 + 2z_1^2 - w^2 .
\end{aligned}
$$

We could now set $z_2 = \sqrt{2} z_1$ and we would get $x_2^2 + z_2^2 - y_2^2 - w^2 = 0$ which is (except for the change of the coordinate names) the equation of Q_2. But this last step is actually redundant, since we can already read the signs from the previous equation. The projective mapping, which transforms the given quadric into the standard form, results from the linear substitutions $(a), \ldots, (e)$ which we have done:

$$
\begin{aligned}
z &\overset{(e)}{=} z_1 + w\,, \\
y &\overset{(c)}{=} y_1 + 2z \overset{(d)}{=} y_2 - w + 2z = y_2 - w + 2(z_1 + w) \\
&= y_2 + 2z_1 + w\,, \\
x &\overset{(a)}{=} x_1 + y \overset{(b)}{=} x_2 - 2z + y = x_2 - 2(z_1 + w) + y_2 + 2z_1 + w \\
&= x_2 + y_2 - w.
\end{aligned}
$$

The order of the variables is of course arbitrary and can be changed. It can also happen that a mixed term occurs without the associated squares, e.g. xy, but neither x^2 nor y^2; in this case $x = u + v$, $y = u - v$ must be substituted.

2. *Elementary row and column transformation:* To do this, we must first write the quadratic form in the form $q(v) = v^T A v$ (with $v^T = (x, y, z, w)$) for some symmetric matrix $A = (a_{ij})$. The coefficients before the squares in $q(x)$ are the diagonal elements, the non-diagonal elements are the half coefficients of the

mixed terms (e.g. $4xz = 2xz + 2zx$ gives $a_{13} = a_{31} = 2$). So we get the matrix

$$A = \begin{pmatrix} 1 & -1 & 2 & 0 \\ -1 & 0 & 0 & -1 \\ 2 & 0 & 2 & 0 \\ 0 & -1 & 0 & 0 \end{pmatrix} \tag{3.20}$$

Now we apply a substitution $v = S\tilde{v}$, hence $q(v) = q(S\tilde{v}) = \tilde{v}^T S^T A S v$. Here S is a concatenation of *elementary matrices*,[32] which are chosen in such a way that at the end $S^T A S$ has a diagonal shape. After each elementary row transformation the corresponding column transformation is to be executed (which together correspond to the transformation $A \mapsto S^T A S$ for an elementary matrix S). If we restrict ourselves to type II, the result of this latter (column) transformation is only that the matrix becomes symmetric again; all coefficients on and below the diagonal remain unchanged.[33] Therefore, we can always perform row and column transformations simultaneously: We perform the row transformation (e.g., adding the first row to the second), but list from the new matrix only the coefficients on and below the diagonal and complete it to a symmetric matrix. This procedure is only a (shorthand) reformulation of the previous one by quadratic completion.[34] We therefore obtain the following scheme:

$$\begin{array}{cccc|cccc|cccc|cccc}
1 & -1 & 2 & 0 & 1 & 0 & 0 & 0 & 1 & 0 & 0 & 0 & 1 & 0 & 0 & 0 \\
-1 & 0 & 0 & -1 & 0 & -1 & 2 & -1 & 0 & -1 & 0 & 0 & 0 & -1 & 0 & 0 \\
2 & 0 & 2 & 0 & 0 & 2 & -2 & 0 & 0 & 0 & 2 & -2 & 0 & 0 & 2 & 0 \\
0 & -1 & 0 & 0 & 0 & -1 & 0 & 0 & 0 & 0 & -2 & 1 & 0 & 0 & 0 & -1
\end{array}$$

[32] Elementary row transformations are: (I) swapping rows, (II) adding a multiple of another row, (III) multiplying a row by a nonzero factor. Elementary column transformations are defined accordingly. Elementary row or column transformations of a matrix A can be accomplished by multiplying A by corresponding matrices S from the left or from the right, respectively; these matrices are called *elementary matrices*. The elementary matrices of type I are *permutation matrices*, which permute the standard basis elements $e_1, \ldots, e_{n+1}$ of $\mathbb{R}^{n+1}$ (e.g. $S = \left(\begin{smallmatrix} & 1 \\ 1 & \end{smallmatrix}\right)$), those of type III are diagonal matrices with all ones except one entry $\lambda \neq 0$, e.g. $S = \left(\begin{smallmatrix} 1 & \\ & \lambda \end{smallmatrix}\right)$, and those of type II have all ones on the diagonal and one more coefficient $\lambda \neq 0$, e. g. $S = \left(\begin{smallmatrix} 1 & \\ \lambda & 1 \end{smallmatrix}\right)$ or $S = \left(\begin{smallmatrix} 1 & \lambda \\ & 1 \end{smallmatrix}\right)$. Namely, if A consists of rows a and b such that $A = \left(\begin{smallmatrix} a \\ b \end{smallmatrix}\right)$, then $\left(\begin{smallmatrix} 1 & \\ \lambda & 1 \end{smallmatrix}\right)\left(\begin{smallmatrix} a \\ b \end{smallmatrix}\right) = \left(\begin{smallmatrix} a \\ b+\lambda a \end{smallmatrix}\right)$, and if A has columns u, v such that $A = (u, v)$, then $(u, v)\left(\begin{smallmatrix} 1 & \lambda \\ & 1 \end{smallmatrix}\right) = (u, v + \lambda u)$.

[33] The coefficients below the diagonal are unchanged by right multiplication with an upper triangular matrix with ones on the diagonal. The coefficients on the diagonal are not changed by the type II transformation because the coefficient of A involved has already been made zero by the preceding row transformation. Example: If $A = \left(\begin{smallmatrix} 1 & -1 \\ -1 & 0 \end{smallmatrix}\right)$ and $S = \left(\begin{smallmatrix} 1 & 1 \\ & 1 \end{smallmatrix}\right)$ then $S^T A = \left(\begin{smallmatrix} 1 & \\ 1 & 1 \end{smallmatrix}\right)\left(\begin{smallmatrix} 1 & -1 \\ -1 & 0 \end{smallmatrix}\right) = \left(\begin{smallmatrix} 1 & -1 \\ 0 & -1 \end{smallmatrix}\right)$ and $S^T A S = \left(\begin{smallmatrix} 1 & -1 \\ 0 & -1 \end{smallmatrix}\right)\left(\begin{smallmatrix} 1 & 1 \\ & 1 \end{smallmatrix}\right) = \left(\begin{smallmatrix} 1 & 0 \\ 0 & -1 \end{smallmatrix}\right)$.

[34] In general, one has to obey the special case that at some stage of the procedure all diagonal elements vanish. Then one must first create a non-vanishing diagonal element by adding another row, and then perform the corresponding column transformation, which in this case may also change the coefficients below the diagonal.

3.10 The Cross-Ratio

In affine geometry we have seen (Sect. 2.5) that the ratio of line segments or vectors with parallel directions remains unchanged under affine mappings: If x, y, z are collinear points in an affine space, the vectors $x-z$ and $y-z$ are linearly dependent, $x-z=\lambda(y-z)$ with $\lambda \in \mathbb{K}$, and the *ratio* of the three points,

$$r(x,y,z)=\frac{x-z}{y-z}=\lambda$$

is invariant under any affine mapping F, that is $r(Fx,Fy,Fz)=r(x,y,z)$ for all collinear point triples x, y, z.

Does something similar hold in projective geometry? Given three distinct collinear points $[x],[y],[z] \in \mathbb{P}^n$, the vectors $x,y,z \in \mathbb{K}^{n+1}$ lie in a two-dimensional subspace $E \subset \mathbb{K}^{n+1}$, and we can choose them to lie on a common affine straight line $g \subset E$, that is

$$x-z=\lambda(y-z) \tag{3.21}$$

for some $\lambda \in \mathbb{K}$. Then the ratio $r(x,y,z)=\lambda$ is defined as before. However, the affine straight line g, on which the vectors x, y, z lie, is not uniquely determined; instead we could use equivalent vectors

$$x'=\alpha x,\quad y'=\beta y,\quad z'=\gamma z \tag{3.22}$$

which lie on another affine straight line g' (see figure):

$$x'-z'=\lambda'(y'-z'). \tag{3.23}$$

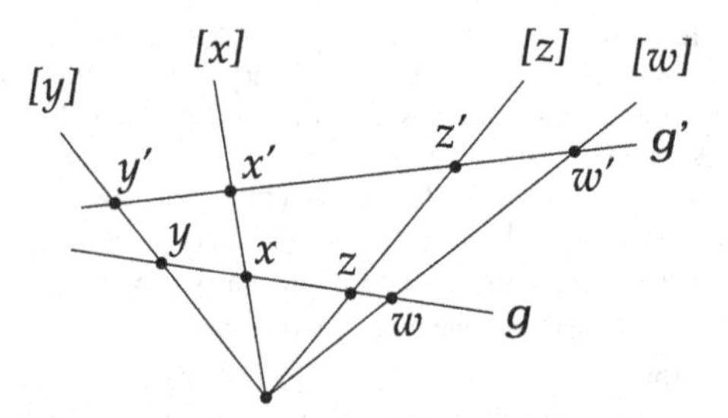

What is the relationship between the two ratios λ and λ'? Inserting (3.22) in (3.23) we get

$$\alpha x=x'=\lambda' y'+(1-\lambda')z'=\lambda'\beta y+(1-\lambda')\gamma z. \tag{3.24}$$

On the other hand, we multiply (3.21) with α and obtain

$$\alpha x = \alpha\lambda y + \alpha(1-\lambda)z. \tag{3.25}$$

Since the vectors y and z are linearly independent, we can use the right-hand sides of (3.24) and (3.25) to compare the coefficients. In particular we see $\lambda'\beta = \alpha\lambda$, thus

$$\lambda' = \frac{\alpha}{\beta}\cdot\lambda. \tag{3.26}$$

The ratio on the straight line g' is therefore not the same as that on the straight line g, but differs by the factor $\frac{\alpha}{\beta}$. Since we have no way to make a choice between g and g', the ratio $\lambda = r(x, y, z)$ cannot be defined in projective geometry.

But still this case is not completely hopeless, because the quotient of the two ratios, $\frac{\lambda'}{\lambda} = \frac{\alpha}{\beta}$ does not depend on z and z' ! Therefore, if we define a fourth collinear point $[w]$ with representatives $w \in g$ and $w' \in g'$, then the new ratios $\mu = \frac{x-w}{y-w}$ and $\mu' = \frac{x'-w'}{y'-w'}$ are related in exactly the same way:

$$\mu' = \frac{\alpha}{\beta}\cdot\mu. \tag{3.27}$$

So $\frac{\lambda'}{\lambda} = \frac{\mu'}{\mu}$ and hence

$$\frac{\lambda'}{\mu'} = \frac{\lambda}{\mu}. \tag{3.28}$$

The quotient $\frac{\lambda}{\mu}$ is therefore the same for each choice of the straight line g and thus it depends only on the homogeneous vectors $[x], [y], [z], [w]$. It is called the *cross-ratio* of the four collinear points, $([x], [y]; [z], [w])$ or shorter $(x, y; z, w)$:

$$(x, y; z, w) = \frac{r(x, y, z)}{r(x, y, w)} = \frac{x-z}{y-z}\cdot\frac{y-w}{x-w}, \tag{3.29}$$

where the four vectors $x, y, z, w \in \mathbb{K}^{n+1}$ are to be chosen on a common affine straight line. The cross-ratio is obviously invariant under projective mappings of $\mathbb{P}^n$ because these come from linear mappings of $\mathbb{K}^{n+1}$.

This notion also applies to $\mathbb{P}^1 = \mathbb{K}\cup\{\infty\}$; then the cross-ratio is simply a double fraction of numbers, among which however also ∞ can occur. For example $\frac{x-\infty}{y-\infty} = 1$ as you can see by replacing ∞ by $1/t$ and then passing to $t = 0$:

$$\frac{x-\infty}{y-\infty} = \left.\frac{x-1/t}{y-1/t}\right|_{t=0} = \left.\frac{xt-1}{yt-1}\right|_{t=0} = 1. \tag{3.30}$$

On $\mathbb{P}^1$ one cannot speak of "straight" mappings (preserving lines); this is one of the already mentioned difficulties in low dimensions (cf. Footnote 5 in Chap. 2). But with the cross-ratio we have a possibility to characterize projective mappings on $\mathbb{P}^1$ geometrically:

Theorem 3.8 *The projective mappings on $\mathbb{P}^1$ are exactly the bijections that preserve the cross-ratio.*

Proof Since projective mappings on $\mathbb{P}^1$ arise from linear isomorphisms on $\mathbb{K}^2$, they obviously preserve the cross-ratio. If now an arbitrary bijection $F : \mathbb{P}^1 \to \mathbb{P}^1$ is given which preserves the cross-ratio, then by composing with a projective mapping we may assume that $F(\infty) = \infty$. This modified F preserves the cross-ratio and the point ∞, hence it preserves

$$(x, y; z, \infty) = \frac{x-z}{y-z} \cdot \frac{y-\infty}{x-\infty} = \frac{x-z}{y-z} = r(x, y, z).$$

Therefore $F|_{\mathbb{A}^1}$ is affine, see Sect. 2.5, and thus the new map F and also the old one are projective. □

An Application: The Complete Quadrilateral

Since all planar quadrilaterals are projectively equivalent, we can construct an arbitrary quadrilateral with its associated diagonals and "centerlines" in $\mathbb{P}^2$ (*complete quadrilateral*) by a projective mapping onto the square in the affine plane $\mathbb{A}^2$ as shown in the right figure, transforming the dotted line SC into the line at infinity:

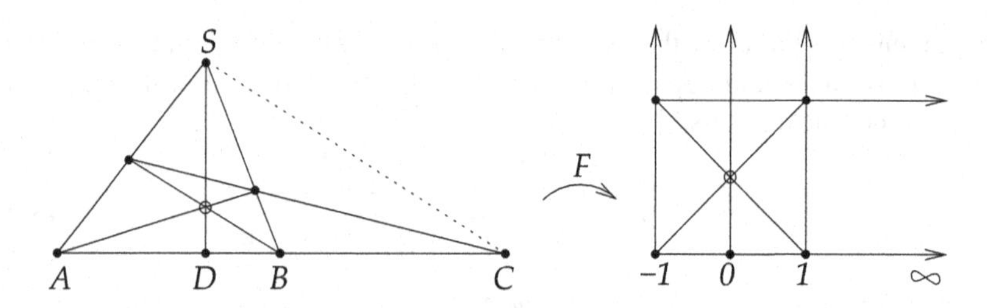

The points A, B, C, D (the *"base points"* of the complete quadrilateral) are thereby mapped to the points $-1, 1, \infty, 0 \in \hat{\mathbb{K}} = \mathbb{P}^1 \subset \mathbb{P}^2$, and we obtain

$$(A, B; C, D) = (-1, 1; \infty, 0) \overset{(3.30)}{=} \frac{1-0}{-1-0} = -1. \tag{3.31}$$

Thus the result is $\frac{A-C}{B-C} \cdot \frac{B-D}{A-D} = (A, B; C, D) = -1$ and hence

$$\frac{A-C}{B-C} = -\frac{A-D}{B-D}. \tag{3.32}$$

The points C and D thus divide the line segment $\overline{AB}$ both outside and inside by the same ratio: $\frac{|A-C|}{|B-C|} = \frac{|A-D|}{|B-D|}$; this is called *harmonic division*. In the figure below the case $\frac{|A-C|}{|B-C|} = 2$ with an associated complete quadrilateral is drawn.

Theorem 3.9 *Four collinear points $A, B, C, D \in \mathbb{P}^2$ are the base points of a complete quadrilateral in $\mathbb{P}^2$ if and only if their division is harmonic, $(A, B; C, D) = -1$.*

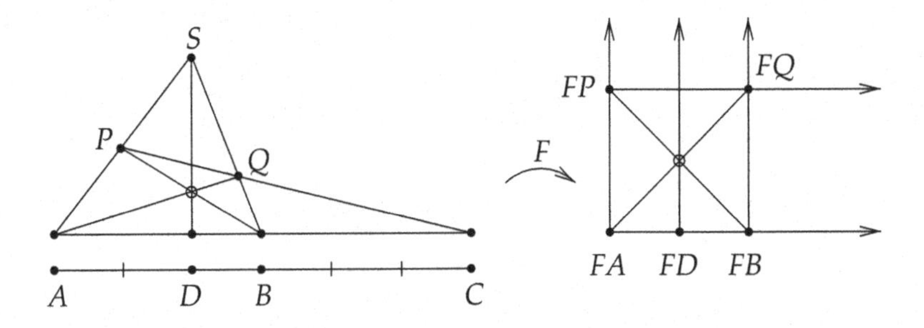

Proof Given are four collinear points A, B, C, D with harmonic division. Since they are collinear, we may assume $A, B, C, D \in \mathbb{P}^1 \subset \mathbb{P}^2$. Because of the harmonic division there is a projective transformation F of $\mathbb{P}^1$ which maps these points to $-1, 1, \infty, 0 \in \mathbb{K} \cup \{\infty\} = \mathbb{P}^1$. In $\mathbb{P}^2$ these are the points $[-1, 0, 1]$, $[1, 0, 1]$, $[1, 0, 0]$, $[0, 0, 1]$. We now choose an arbitrary point $S \in \mathbb{P}^2 \setminus \mathbb{P}^1$ and connect S with A, D, B. Further we choose an arbitrary point $P \neq A, S$ on the straight line AS, connect it with C and call Q the intersection $PC \wedge SB$. Now we extend F to a projective transformation of $\mathbb{P}^2$ which maps P to $[-1, 2, 1]$ and S to the intersection point of the line $FA \vee FP$ with the line at infinity, which is $[0, 1, 0]$. Thus the quadrilateral $ABPQ$ is mapped to a square in $\mathbb{A}^2 \subset \mathbb{P}^2$ and the given figure is a complete quadrilateral, i.e. the intersection point of the diagonals, $AQ \wedge BP$, lies on the "center line" SD. □

4 Distance: Euclidean Geometry

Abstract

In the third century BC, Euclid compiled the mathematical knowledge of the time in his book "The Elements". In his geometrical considerations, measurements play a role from the very beginning: distances, angles, areas, volumes. Distance is the fundamental concept. Similarly as before, we do not want to define this concept axiomatically, but derive it from intuitive considerations and only then incorporate it into our framework of linear algebra, namely with the help of the scalar product. The basis for this is the Pythagorean theorem. We will study the group of structure-preserving transformations, the "isometries", in much more detail this time; their discrete and finite subgroups will also be considered. These, in turn, have to do with crystals and with the Platonic solids; we also introduce the latter in higher dimensions. Next to straight lines, the simplest entities of plane geometry are conic sections, which we will now also examine in terms of lengths and distances. Interestingly enough, this problem becomes more accessible to intuition if we take the term "conic section" literally and also consider the cone in space.

4.1 The Pythagorean Theorem

The theorem of *Pythagoras* is the most important theorem concerning the notion of distance, since it can be used to calculate the distance between two points with given (rectangular) coordinates. However, it did not originate from Pythagoras at all, but was probably known long before to the Egyptians, Indians and Babylonians.[1] It reads:

---

[1] https://en.wikipedia.org/wiki/Pythagorean_theorem, https://link.springer.com/article/10.1057/jt.2009.16.

J.-H. Eschenburg, *Geometry – Intuition and Concepts*,
https://doi.org/10.1007/978-3-658-38640-5_4

In a rectangular ("right") triangle with catheti a, b (= sides that are adjacent to the right angle) and hypotenuse c (= side opposite to the right angle) the three side lengths satisfy $a^2 + b^2 = c^2$.

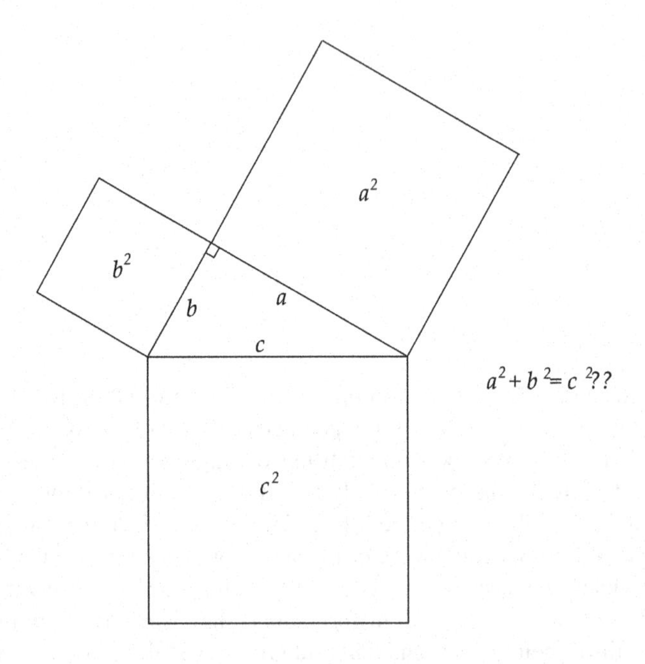

As before, we begin by assuming that we are completely familiar with the intuitive planar geometry. In particular, we know not only what points, straight lines and incidence mean, we also know the meaning of notions like length, distance, angle, area. However, the figure above does not show in any way why the areas of the two small squares together should be equal to the large square area. As always in geometry, a *construction* is needed to trace the hidden back to the obvious. The origin of one of the earliest of these constructions (see figure) is probably India.

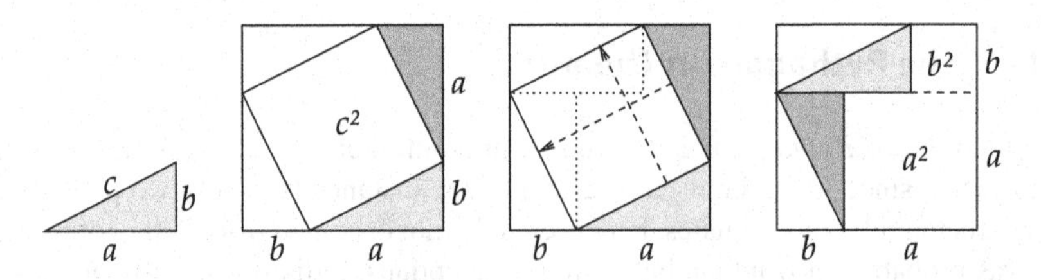

Euclid[2] gave a different proof in his "Elements," the ancient mathematics compendium by which geometry was taught as late as the nineteenth century.

[2] Euclid of Alexandria, about 325–265 BC.

Euclid's proof has found stronger expression in textbooks, though it is more complicated. One decomposes c into its segments p and q below the sides a and b. This makes the square c^2 decomposed into two rectangles cp and cq. Now we show $a^2 = cp$ (and correspondingly $b^2 = cq$). To prove this, the rectangle cp is transformed into a parallelogram by shearing, with no change in area (the triangle cut off at the bottom is added back at the top), this parallelogram is rotated by 90° and then transformed by a second shearing[3] into the square over a.

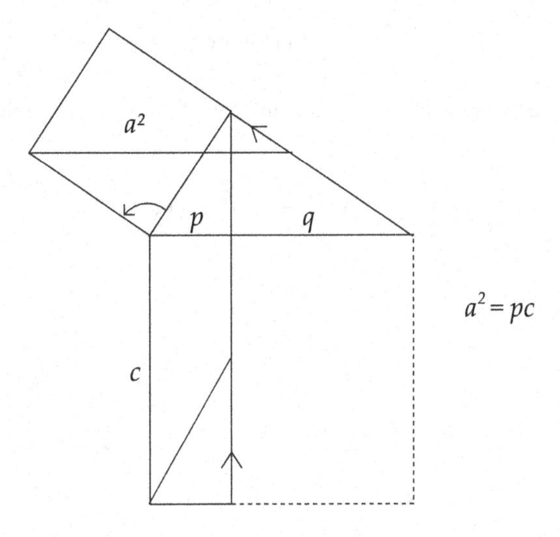

A third proof uses the *similarity* (equal shape but possibly different size) of the given rectangular triangle Δ_c with hypotenuse c to the smaller rectangular triangles Δ_a, Δ_b with hypotenuses a and b, which are formed when decomposing the triangle Δ_c by its altitude. So all three triangles are similar to a rectangular triangle Δ_1 with hypotenuse one and some area F and arise (except for rotations and reflections) from Δ_1 by homotheties with the scaling factors a, b and c. Then the area is multiplied by the square of the scaling factor, and because the small triangles decompose the large one, the assertion follows:

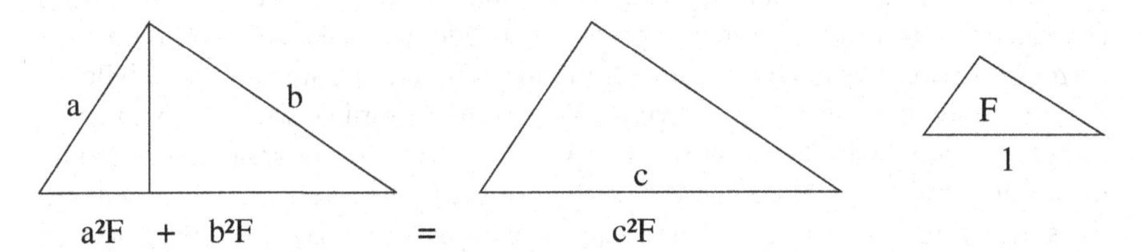

[3] In the second shear, we need to cut the parallelogram and the square into strips to see the equality of area by cutting and attaching.

Although it is a theorem about *side lengths*, all three proofs use the *area* and its transformation properties which are known from everyday life: invariance under decompositions and rotations (see the subsequent remark).

If we now include the concept "distance in the plane" into our axiomatic framework of the vector space $\mathbb{R}^2$ we proceed conversely: We *define* the *norm* or *length* $|x|$ of a vector $x = (x_1, x_2) \in \mathbb{R}^2$, that is the distance of the point x from the origin o, using Pythagoras' formula:

$$|x|^2 = (x_1)^2 + (x_2)^2. \tag{4.1}$$

For the length of a vector $x = (x_1, x_2, x_3)$ in space $\mathbb{R}^3$ we get an analogous formula:

$$|x|^2 = (x_1)^2 + (x_2)^2 + (x_3)^2. \tag{4.2}$$

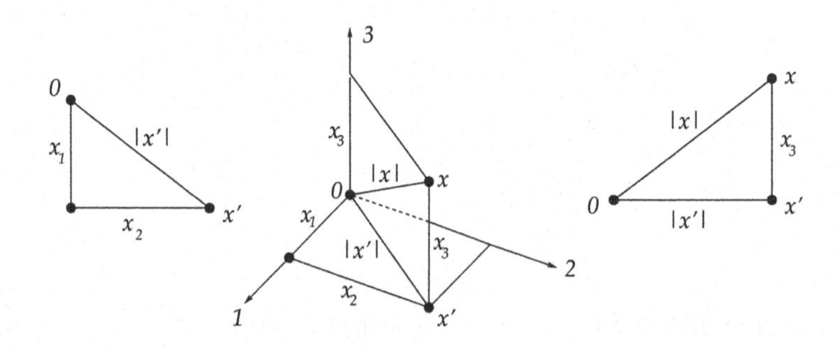

This follows in two steps by considering successively two planes: the x_1x_2-plane with the projection $x' = (x_1, x_2, 0)$ (left figure) and the plane containing x' and the x_3-axis (right figure). By applying the Pythagorean theorem twice we find $|x'|^2 = x_1^2 + x_2^2$ as well as $|x|^2 = |x'|^2 + x_3^2$ and thus the assertion (4.2).

Remark We have used for the three proofs of the Pythagorean theorem the properties of the area of plane figures, in particular the invariance under translations and rotations. But there is a big difference between translations and rotations: the *area* of a figure is defined by counting the paraxial unit squares (edges parallel to x- and y-axis) that cover it (subdividing these squares further if necessary in order to better approximate the figure). Then the invariance of the area under parallel displacements and under 90-degree-rotations of the figure is obvious, because these transformations convert paraxial unit-squares into just such ones. But the invariance under rotations by angles $< 90°$ is not self-evident, because the geometry of the covering unit squares is now completely different (left and middle figure below), so it is not at all immediately clear that their numbers always agree.

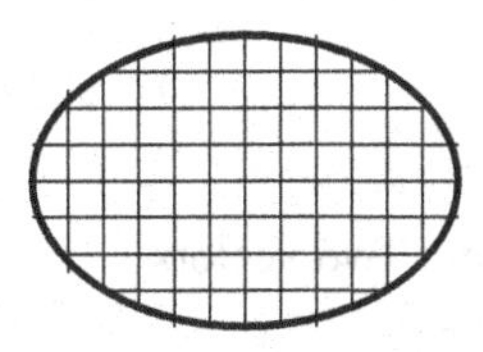
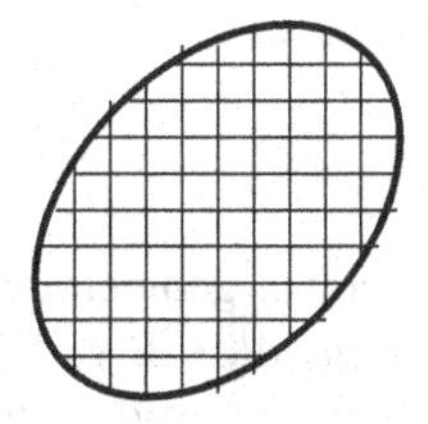
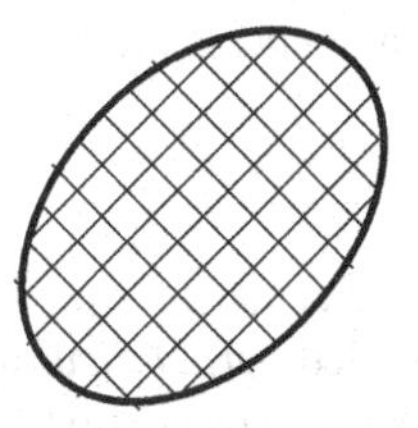

The first and the third proof, however, use the general invariance under rotations, only the second (by the master Euclid) avoids it. To fill this gap, first rotate the figure together with the covering squares; the rotated figure is now covered by squares that are also rotated, and no longer paraxial (right figure); their number is obviously still the same as that of the paraxial ones that previously covered the unrotated (left) figure. To show the equality of the areas of the original and the rotated figure, one needs only to convince oneself that the unit squares did not change their area during the rotation. To see this, use the circle as a norm figure; it remains unchanged under rotations and is covered by as many rotated as paraxial unit squares, so the areas of those are equal.

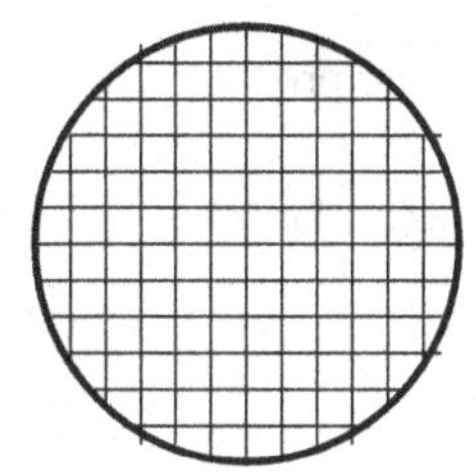
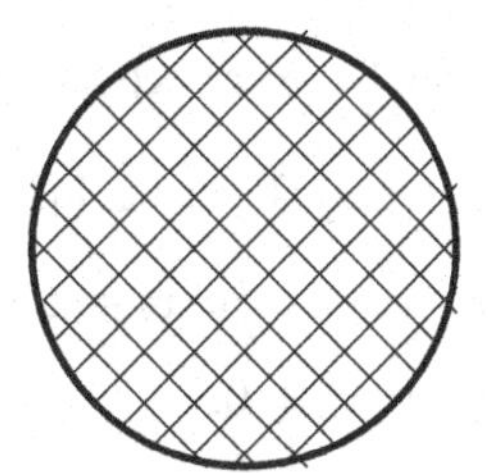

4.2 The Scalar Product in $\mathbb{R}^n$

We now want to transfer the geometry of the distance to arbitrary dimensions n. For this we define the *norm* or *length* analogously for a vector $x = (x_1, \ldots, x_n) \in \mathbb{R}^n$,

$$|x|^2 = x_1^2 + \cdots + x_n^2. \tag{4.3}$$

As *distance* of any two points $x, y \in \mathbb{R}^n$ we define (as in the visual dimensions $n \leq 3$) the length of the difference vector, $|x - y|$. With the help of the *scalar product* or *inner product* that assigns to any two vectors $x, y \in \mathbb{R}^n$ the real number[4]

$$\langle x, y \rangle := x^T y = x_1 y_1 + \ldots + x_n y_n \in \mathbb{R}, \tag{4.4}$$

[4] In general, we think of a vector $x \in \mathbb{R}^n$ as a column rather than a row, so that the rule "row times column" from the matrix calculus works. The transpose of a vector, x^T, is thus a row, and $x^T y$ (row times column) is a real number.

we can express the norm (4.3) simply as

$$|x|^2 = \langle x, x \rangle. \tag{4.5}$$

But the scalar product has yet another geometric meaning: Two vectors $x, y \in \mathbb{R}^n$ are called *perpendicular* or *orthogonal*, $x \perp y$, if $\langle x, y \rangle = 0$.

Definitions are up to us, but here an already existing term from our everyday life, "perpendicular", is turned into a mathematical definition. Is this consistent with our experience? From intuitive geometry we know when two vectors (straight segments) x and y are perpendicular, $x \perp y$, namely when the distances from y to x and to $-x$ are equal, $|y - x| = |y - (-x)| = |y + x|$ as in the figure on the left:

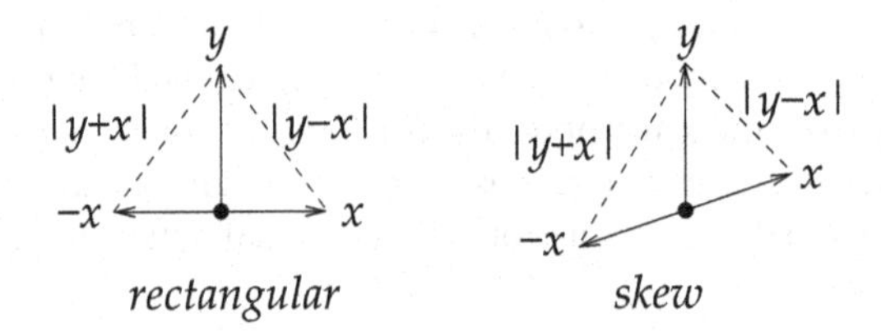

This property can be expressed using the scalar product:

$$\begin{aligned} x \perp y &\iff |y+x|^2 = |y-x|^2 \iff \langle y+x, y+x \rangle = \langle y-x, y-x \rangle \\ &\iff \langle y, y \rangle + \langle x, x \rangle + 2\langle y, x \rangle = \langle y, y \rangle + \langle x, x \rangle - 2\langle y, x \rangle \\ &\iff \langle y, x \rangle = 0. \end{aligned}$$

Therefore, our everyday experience entitles us to consider two vectors $x, y \in \mathbb{R}^n$ *perpendicular* or *orthogonal* if $\langle x, y \rangle = 0$.

For any rectangular triangle $(0, x, y)$ (with $x, y \in \mathbb{R}^n$ and $x \perp y$), the Pythagorean theorem now can be derived in our framework (no wonder, since we have already built it into the definition of distance):

$$|x - y|^2 = \langle x - y, x - y \rangle = \langle x, x \rangle + \langle y, y \rangle - 2\langle x, y \rangle = |x|^2 + |y|^2.$$

Therefore any two-dimensional subspace $E \subset \mathbb{R}^n$ ("plane"), spanned by two linearly independent vectors $x, y \in \mathbb{R}^n$ (that is, $\{x, y\}$ is a basis of E), has the "same" geometry as the intuitive plane. In this sense we can imagine the n-dimensional space $\mathbb{R}^n$ by intuitive geometry, as we also understand the three-dimensional space geometrically better by considering suitable planes, as in the figure after Eq. (4.2).

More generally, with the help of the scalar product also the *angle* α between any two vectors $x, y \in \mathbb{R}^n$ can be defined: If $x \neq 0$, one can decompose y into components parallel and perpendicular to x,

$$y^{||} = \frac{1}{|x|^2} \langle y, x \rangle x\,, \quad y^{\perp} = y - y^{||}. \tag{4.6}$$

In fact, $y^{||}$ is a scalar multiple of x, and $y^{\perp}$ is perpendicular to x, because according to (4.6) $\langle y^{||}, x\rangle = \langle y, x\rangle$ and thus $\langle y^{\perp}, x\rangle = \langle y, x\rangle - \langle y^{||}, x\rangle = 0$.

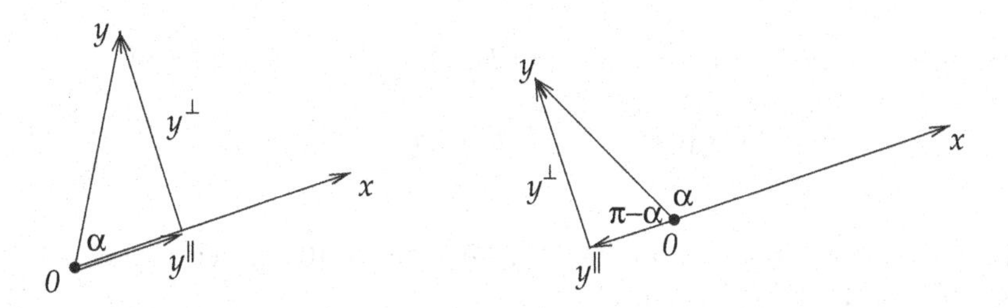

Let us apply the familiar "school definition" of cosine and sine to the rectangular triangle with the vertices 0, y, $y^{||}$,

cosine = adjacent cathetus/hypotenuse,
sine = opposite cathetus/hypotenuse.

Then from the left and right figures we obtain using $\cos(\pi - \alpha) = -\cos\alpha$:

$$\cos\alpha = \pm\frac{|y^{||}|}{|y|} = \frac{\langle y, x\rangle}{|x||y|}. \tag{4.7}$$

In particular, this quotient has absolute value ≤ 1 because according to Pythagoras $|y| = \sqrt{|y^{||}|^2 + |y^{\perp}|^2} \leq |y^{||}|$, and equality holds precisely when $y^{\perp} = 0$, that is when x, y are linearly dependent. This is the *Cauchy-Schwarz inequality*,

$$\langle x, y\rangle \overset{(1)}{\leq} |\langle x, y\rangle| \overset{(2)}{\leq} |x||y| \tag{4.8}$$

with equality at (1) if and only if $\langle x, y\rangle \geq 0$ (i.e. $\alpha \leq \pi/2$) and at (2) precisely if x, y are linearly dependent.

From the Cauchy-Schwarz inequality (4.8) also the *triangle inequality* follows,

$$|x + y| \leq |x| + |y|, \tag{4.9}$$

for $(|x| + |y|)^2 - |x + y|^2 = 2|x||y| - 2\langle x, y\rangle \geq 0$, and equality holds exactly when $\langle x, y\rangle = |x||y|$, that is, when x, y point in the same direction, $y = \mu x$ for a $\mu \geq 0$. The geometrical interpretation is: The sum of two triangle sides is always greater than the third, i.e. for three points $a, b, c \in \mathbb{R}^n$ always holds

$$|b - a| \leq |b - c| + |c - a|, \tag{4.10}$$

"detours are longer". We simply set $x = b - c$ and $y = c - a$ in (4.9). Here the equality discussion is particularly important:

Theorem 4.1 *For every two points $a, b \in \mathbb{R}^n$ holds: The line segment*

$$[a, b] = \{a + \lambda(b - a);\ \lambda \in [0, 1]\}$$

consists exactly of the points $c \in \mathbb{R}^n$ for which equality holds in (4.10):

$$[a, b] = \{c \in \mathbb{R}^n;\ |a - b| = |a - c| + |c - b|\}\,. \tag{4.11}$$

Proof If $c = a+\lambda(b-a) = b+(1-\lambda)(a-b)$ with $\lambda \in [0, 1]$, then $c-a = \lambda(b-a)$ and $b - c = (1 - \lambda)(b - a)$, hence

$$|a - c| + |c - b| = (\lambda + 1 - \lambda)|a - b| = |a - b|.$$

Conversely, if the equality in (4.10) holds, i.e. $|x + y| = |x| + |y|$ for $x = b - c$ and $y = c - a$, then $y = \mu x$ for a $\mu \geq 0$ (discussion of equality of (4.9)), so $c - a = \mu(b - c)$ and thus $c = \frac{1}{1+\mu} a + \frac{\mu}{1+\mu} b \in [a, b]$ because the numbers $\frac{1}{1+\mu}$ and $\frac{\mu}{1+\mu}$ are non-negative and add up to one. □

$\mathbb{R}^n$ with the standard scalar product forms the exact mathematical model for Euclidean geometry. However, the origin $0 \in \mathbb{R}^n$ has no special geometric meaning; therefore, we must actually proceed to the associated affine space. Also, the distinction of the standard basis $e_1 = (1, 0, \dots, 0), \dots, e_n = (0, \dots, 0, 1)$ is not justified geometrically; any other orthonormal basis is geometrically perfectly equivalent. We may, therefore, just as well consider an arbitrary n-dimensional vector space V over $\mathbb{R}$ with a general scalar product $\langle\ ,\ \rangle$;[5] by choosing any orthonormal basis,[6] we can identify V with $\mathbb{R}^n$ equipped with the standard scalar product. Therefore, such a vector space with scalar product is also called a *Euclidean vector space.* For example, any subspace of $\mathbb{R}^n$ with the restricted standard scalar product is itself a Euclidean vector space; in particular, any two-dimensional subspace of the $\mathbb{R}^n$ is an exact model of "the" Euclidean plane which is why the geometry of a high-dimensional $\mathbb{R}^n$ is in large parts still amenable to visualization. For many applications, one can even dispense with the finiteness of the dimension, as is the case in functional analysis (space of L^2-functions with the scalar product $\langle f, g\rangle = \int(fg)$) and in physics (space of states in quantum mechanics).

It is not so surprising why the Pythagorean theorem was first proved using area transformations: We are, after all, talking about squares, and these can be visualized as areas. Euclidean geometry is closely connected with a quadratic form q, the sum of squares $q(x) = \sum_i (x_i)^2$, because the Euclidean norm is $|x| = q(x)^{1/2}$. Since nineteenth century, mathematicians have also discussed other norms and

[5] $V \times V \to \mathbb{R}$ bilinear, symmetric ($\langle v, w\rangle = \langle w, v\rangle$), positive definite ($\langle v, v\rangle > 0$ for $v \neq 0$).

[6] A basis $b_1, \dots, b_n$ of V is called *orthonormal* if $\langle b_i, b_j\rangle = \delta_{ij} = \begin{cases} 1 \text{ for } i = j \\ 0 \text{ for } i \neq j \end{cases}$.

notions of distance, e.g. the *p-norm* $|x|_p = (\sum_i (x_i)^p)^{1/p}$ for arbitrary $p > 0$. Interestingly, only for $p = 2$, i.e. in the Euclidean case, there exists a large group of distance-preserving transformations (*isometries*), which was first proved by *H. von Helmholtz*.[7] This is related to a more general observation: Bilinear forms (2-linear forms) have a large automorphism group (the group of linear transformations keeping the 2-linear form invariant), but the automorphism group of a p-linear form for $p \geq 3$ is large only in a few exceptional cases.[8]

Therefore, for any number of dimensions (according to Tits) there are only three types of "geometries": They belong to a vector space with either no further structure or with a non-degenerate symmetric or antisymmetric bilinear form. These are *projective*, *polar* and *symplectic* geometry. In this sense, metric geometry belongs to the polar geometry. Later, we shall proceed with the geometry of circles and spheres (Sects. 6.5 and 6.6) which are further representatives of polar geometry. Symplectic geometry underlies Hamiltonian mechanics and is at present a very active field of research; in this book, however, it will play no further role. Besides these *classical geometries*, there exist in certain dimensions, e.g. in dimension 7, certain *exceptional geometries*, which are all somehow connected with the above mentioned octonion algebra $\mathbb{O}$.

4.3 Isometries of Euclidean Space

We recall that a linear mapping or matrix $A : \mathbb{R}^n \to \mathbb{R}^n$ is called *orthogonal*[9] if its columns $Ae_1, \ldots, Ae_n$ form an orthonormal basis: $\delta_{ij} = \langle Ae_i, Ae_j\rangle = e_i^T(A^TA)e_j$, or $A^TA = I$ (where $I = (\delta_{ij})$ denotes the unit matrix on $\mathbb{R}^n$). These are exactly the linear maps which preserve the scalar product: $\langle Ax, Ay\rangle = \langle x, y\rangle$ for all $x, y \in \mathbb{R}^n$, for $\langle Ax, Ay\rangle = (Ax)^TAy = x^TA^TAy$. Of course, in particular, the norm is preserved: $|Ax|^2 = \langle Ax, Ax\rangle = \langle x, x\rangle = |x|^2$, but the converse also holds: If A preserves the norm, $|Ax| = |x|$ for all $x \in \mathbb{R}^n$, then A is already orthogonal. This results from the polarization trick:

$$2\langle Ax, Ay\rangle = |A(x+y)|^2 - |Ax|^2 - |Ay|^2 = |x+y|^2 - |x|^2 - |y|^2 = 2\langle x, y\rangle.$$

Since these properties are preserved under composition and inversion, the orthogonal matrices form a subgroup $O(n)$ of the group $GL(\mathbb{R}^n)$ of all invertible matrices, called the *orthogonal group*.

[7] Hermann von Helmholtz, 1821 (Potsdam)–1894 (Berlin), *Über die Tatsachen, welche der Geometrie zugrunde liegen*, Nachr. der Ges. d. Wiss. Göttingen 1868, https://www.e-rara.ch/zut/content/titleinfo/1309290; see also Hermann Weyl: *Space, Time, Matter*, Springer 1919, English: Dover 1952.

[8] For example, in $\mathbb{R}^7$ there is an alternating 3-linear form with a large group of automorphisms; it is the automorphism group of the previously mentioned *octonion algebra*.

[9] It should better be called *orthonormal*, but "orthogonal" has become common.

Theorem 4.2 *Distance preserving mappings (isometries) of the Euclidean space $\mathbb{R}^n$ are affine mappings, and an affine mapping $x \mapsto Ax + b$ is an isometry if and only if $A \in O(n)$.*

Proof By Theorem 4.1, line segments are characterized by a purely metric property which is preserved under isometries. An isometry $F : \mathbb{R}^n \to \mathbb{R}^n$ thus maps line segments to line segments, and hence straight lines to straight lines. Furthermore, F is also a parallel map, because parallels are straight lines of constant distance from each other. Therefore F is affine, $F(x) = Ax + b$ for a linear mapping A and a vector b. Then $|Fx - Fy| = |Ax + b - Ay - b| = |Ax - Ay| = |A(x - y)|$, and $|A(x - y)| = |x - y|$ for all x, y if and only if $A \in O(n)$. □

The isometries of $\mathbb{R}^n$ thus form a subgroup of the affine group, called the *Euclidean group $E(n)$*; like the affine group it is a *semidirect product* (cf. Exercises 6 and 17) of the orthogonal group $O(n)$ with the translation group $\mathbb{R}^n$. In particular, every isometry F is a composition of an orthogonal map with a translation: $F = T_b \circ A$. If the orthogonal map A has positive determinant ($A \in SO(n)$)[10] and thus transforms oriented bases back into oriented bases,[11] then F is called an *oriented isometry* or *proper motion.*

4.4 Classification of Isometries

A special role among the isometries is played by the *hyperplane reflections*. The reflection along an affine hyperplane $H \subset \mathbb{R}^n$ is the isometry S, which fixes the points of H and maps every point on one side of H to its counterpart at the same distance on the other side of H. If H passes through the origin and thus is a linear subspace, we speak of a *linear hyperplane reflection*; then

$$S(x) = x - 2\frac{\langle x, h\rangle}{\langle h, h\rangle}h \tag{4.12}$$

for all x, where h is a nonzero vector perpendicular to H; the component of x perpendicular to H must be subtracted twice to get “to the other side”.

Theorem 4.3 *Any nontrivial orthogonal map on $\mathbb{R}^n$ is the composition of at most n linear hyperplane reflections. Every isometry on $\mathbb{R}^n$ is the composition of at most $n + 1$ hyperplane reflections.*

[10] $\det A = \pm 1$ for any $A \in O(n)$ since $A^T A = I$ and hence $(\det A)^2 = 1$.

[11] Two bases of an n-dimensional $\mathbb{R}$-vector space are called *equally oriented,* if the transition matrix between them has positive determinant. This is an equivalence relation on the set of bases, and there are exactly two equivalence classes. We call the choice of one of the two classes an *orientation* of the vector space. A basis belonging to this class is then called an *oriented basis.* In $\mathbb{R}^n$, orientation is given by the class of the standard basis $(e_1, \dots e_n)$.

Proof We show the first statement by induction on n. For $n = 1$ there is only one non-trivial orthogonal mapping, namely $x \mapsto -x$, and this is a reflection along the "hyperplane" $\{0\} \subset \mathbb{R}^1$. Now if $A \in O(n)$ and by chance $Ae_n = e_n$ then the subspace $\mathbb{R}^{n-1} = (e_n)^\perp$ is mapped onto itself and $A' := A|_{\mathbb{R}^{n-1}} \in O(n-1)$. According to the induction assumption, A' is the composition of reflections along k hyperplanes $H_1', \dots, H_k' \subset \mathbb{R}^{n-1}$ with $k \leq n-1$. Then also A is a composition of reflections along k hyperplanes $H_1, \dots, H_k \subset \mathbb{R}^n$ with $H_i = H_i' + \mathbb{R}e_n$.

But if $Ae_n \neq e_n$, we consider $B = SA$ where S is the reflection along the hyperplane $H = (Ae_n - e_n)^\perp$ which is the perpendicular bisector on the line segment $[e_n, Ae_n]$. Then S interchanges the vectors e_n and Ae_n, and so $Be_n = SAe_n = e_n$. As before B is the composition of at most $n-1$ linear hyperplane reflections and thus $A = SB$ is the composition of at most n such reflections. This proves the first statement.

If F now is an arbitrary isometry with $F(0) \neq 0$, we consider the reflection S along the hyperplane H which is the perpendicular bisector on the line segment $[0, F(0)]$ (viz. $H = F(0)^\perp + \frac{1}{2}F(0)$); this reflection interchanges the points 0 and $F(0)$. Then $A = S \circ F$ is linear (rather than affine) because $A(0) = S(F(0)) = 0$, and thus (being an isometry) it is an orthogonal map. So A is the composition of at most n hyperplane reflections, and $F = S \circ A$ is the composition of at most $n+1$ hyperplane reflections. □

Corollary 4.4 *Any nontrivial isometry on $\mathbb{R}^2$ belongs to one of the following four classes:*

(a) translations (by an arbitrary vector),
(b) rotations (about an arbitrary center and with arbitrary rotation angle),
(c) reflections (along an arbitrary axis),
(d) glide reflections (along an arbitrary axis).

Proof An isometry of $\mathbb{R}^2$ can be composition of 1, 2 or 3 reflections. For one reflection we are in case (c). In the case of two reflections, their axes may intersect—then we obtain a rotation about the intersection point of the axes (case (b)) where the angle of rotation is twice the angle between the two axes, or they may be parallel and we get a translation perpendicular to the two axes by twice their distance. In the case of three reflections, all three axes can be parallel, in which case we get a reflection on another axis parallel to them. Otherwise, at most two axes are parallel, and the third intersects the other two. Then we can put the axes in a special position and show that it is a glide reflection (class (d)) as we will show now.

This is based on the following observation: The composition of two reflections on non-parallel axes is a rotation about their intersection point, but we obtain the same rotation if we rotate the pair of axes by an arbitrary angle about the intersection point.

Let the given isometry be $F = S_3S_2S_1$. The axes of the three reflections S_1, S_2, S_3 are correspondingly denoted by 1, 2, 3. We can assume that the axes 1 and 2 have a common point of intersection D: If they should be parallel, then by assumption axes 2 and 3 must intersect, and after rotating this pair of axes about their common point, 1 and 2 intersect; in this case S_2 and S_3 have been changed individually, but not their composition S_3S_2. Now we rotate axes 1 and 2 about their intersection point D (figure (a) below) and obtain a new pair of axes $(1', 2')$ through D, where $2'$ meets the axis 3 at a point D' at a right angle (figure (b)). Then we rotate the pair of axes $(2', 3)$ about D'; the new pair of axes $(2'', 3'')$ has the property that $3''$ is perpendicular to $1'$ and $2''$ parallel to $1'$ (figure (c)). Now $F = S_3S_2S_1 = S_3S_2'S_1' = S_3''S_2''S_1'$, and thus F is the composition of the translation $T = S_2''S_1'$ (the mirror axes $2''$ and $1'$ are parallel to each other) with the reflection S_3'' whose mirror axis is parallel to the translation direction of T. Such a mapping is a *glide reflection.* □

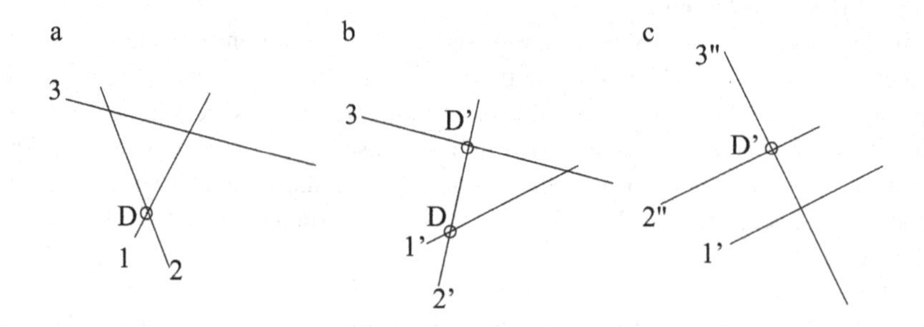

Corollary 4.5 *Every oriented orthogonal map on $\mathbb{R}^3$ has an axis: If $A \in O(3)$ is oriented, i.e.* $\det A > 0$, *then there is a vector $a \neq 0$ with $Aa = a$.*

Proof A can be a composition of 1, 2 or 3 linear plane reflections, but because $\det A > 0$, 1 or 3 plane reflections are not possible. So $A = S_1S_2$ where S_1, S_2 are reflections along two planes H_1, H_2 through 0. These intersect in a straight line $\mathbb{R}a$ which is fixed under S_1 and S_2, hence under A. □

Corollary 4.6 *Any oriented isometry on $\mathbb{R}^3$ is a screw motion, i.e. a rotation followed by a translation in the direction of the axis of rotation.*

Proof Let the given isometry be $F = T_b \circ A$ with $A \in SO(3)$. By Corollary 4.5, A has a fixed vector a. We consider the orthogonal decomposition $\mathbb{R}^3 = \mathbb{R}a \oplus a^\perp$; every vector $x \in \mathbb{R}^3$ splits uniquely into $x = x_a + x_\perp$ with $x_a \in \mathbb{R}a$ and $x_\perp \in a^\perp$; in particular $b = b_a + b_\perp$. Then $Ax = x_a + Ax_\perp$ and $Fx = x_a + b_a + Ax_\perp + b_\perp$. The last two summands $Ax_\perp$ and $b_\perp$ lie in $a^\perp$ and define an isometry $F^\perp$ of the plane $a^\perp$, namely $F^\perp(x_\perp) = Ax_\perp + b_\perp$. This is oriented, but not a translation, so it is a rotation by Corollary 4.4. Let $c \in a^\perp$ be its center of rotation. The isometry F is therefore the composition of a rotation and a translation parallel to the rotation axis. In fact, the rotation is $x = x_a + x_\perp \mapsto x_a + F^\perp x_\perp$ with axis $c + \mathbb{R}a$, and the translation is T_{b_a}. Since b_a is a multiple of a, the translation vector is parallel to the axis. □

Corollary 4.7 *Any isometry on $\mathbb{R}^3$ belongs to one of the following three types:*

(a) a trivial extension of an isometry on $\mathbb{R}^2$,
(b) a screw motion,
(c) a "rotation-reflection": reflection and rotation in the mirror plane.

Proof Sketch According to Corollary 4.6 we only have to study the non-oriented case, i.e. an isometry F consisting of (one or) three reflections. If the three mirrors contain three parallel straight lines, then case (a) happens. Otherwise, two of the three mirrors intersect in a straight line that also meets the third mirror. This intersection point is fixed by all three reflections; it is therefore a fixed point of our isometry F, and we can place it at the origin 0. Now the three mirrors intersect a sphere with center 0 in three great circles. By a process similar to that in the proof of Corollary 4.4 (rotating each two mirrors about their intersection line), we can make the third mirror intersect the other two perpendicularly; the result is a rotation-reflection.

4.5 Platonic Solids

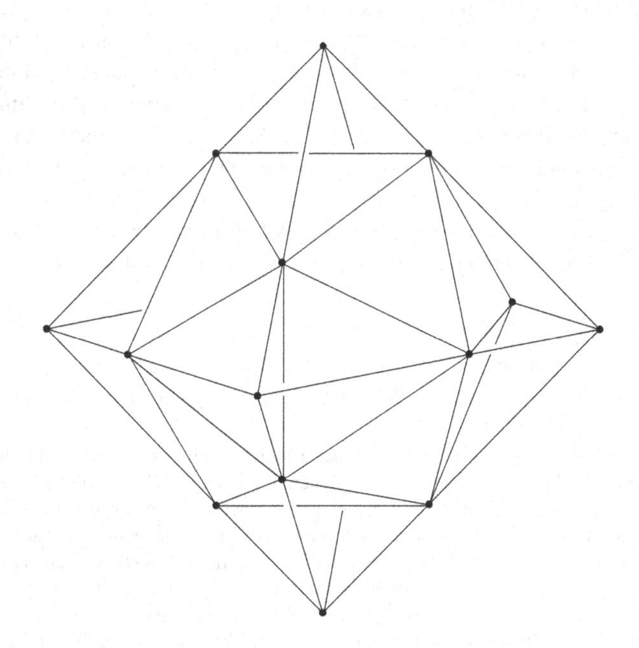

A *Platonic solid*[12] in space $\mathbb{R}^3$ is a compact convex set[13] with non-empty interior, bounded by congruent regular plane polygons, with an equal number of these polygons adjacent to each vertex. There are five such solids: tetrahedron, cube, octahedron, dodecahedron, and icosahedron.[14] The figure shows the octahedron with the vertices $(\pm 1, 0, 0)$, $(0, \pm 1, 0)$, $(0, 0, \pm 1)$ and inside it an icosahedron whose 12 vertices lie on the 12 edges of the octahedron, each edge being subdivided in the golden ratio (see Exercise 10).

Why aren't there more than these five? To understand this, consider a vertex of such a solid with the polygons adjacent to it, the *star* of the vertex; this already determines the solid unambiguously (see below). The interior angles of all polygons that meet there must add up to a value of $< 360°$ so that a convex spatial corner can be created. For regular triangles, the interior angle is $60°$; so in each corner 3, 4 or 5 triangles may meet (tetrahedron, octahedron, icosahedron); 6 triangles already form a plane pattern ($6 \cdot 60 = 360$) and no longer a spatial corner. For regular quadrilaterals (squares) this angle is $90°$; therefore only 3 of them can meet at a corner (cube) because 4 already form a plane pattern ($4 \cdot 90 = 360$). Also with pentagons only 3 can come together (dodecahedron) because the angle is $108°$ and $4 \cdot 108 > 360$. Hexagons do not occur any more, because already three of them form a plane pattern (honeycomb pattern), and higher polygons with angle $> 120°$ are out of the question; so the list of Platonic solids is complete.

Do "Platonic solids" exist in higher dimensions? For this we must first clarify the term: A *Platonic solid in* $\mathbb{R}^n$ is defined as a regular compact convex polytope. These notions need to be explained. An n-dimensional convex *polytope* $P \subset \mathbb{R}^n$ is the intersection of finitely many *half spaces* $H_{v,\lambda} = \{x \in \mathbb{R}^n;\ \langle x, v\rangle \geq \lambda\}$ with $v \in \mathbb{R}^n \setminus \{0\}$ and $\lambda \in \mathbb{R}$, and we also assume that this intersection has non-empty interior. By definition, a polytope is closed; additionally we assume that it is bounded, hence compact. An n-dimensional compact convex polytope is bounded by finitely many $(n-1)$-dimensional compact convex polytopes, the *faces*, whose faces in turn are $(n-2)$-dimensional compact convex polytopes, etc. A descending chain of faces $P \supset S_1 \supset S_2 \supset \cdots \supset S_n$ down to a point S_n (a *vertex* or *corner* of P) will be called a *flag*. The convex polytope is already determined by its vertices; it is their *convex hull*.[15] The *symmetry group* G_P of a polytope P is the set of all isometries

[12] Plato, 427–347 BC (Athens).

[13] A subset $C \subset \mathbb{R}^n$ is called *convex* if for any two points $x, y \in C$ the connecting line segment $[x, y]$ is contained in C.

[14] At the time of Pythagoras probably only three of these solids were known while the remaining two are ascribed to *Theaetetus* (415–365 BC, Athens), a friend of Plato who treats all five in his dialogue "Timaeus". In [7], p. 505, Footnote 1, it is claimed that the two missing solids were the icosahedron and the dodecahedron, while John Baez https://math.ucr.edu/home/baez/icosahedron/ cites a scholium saying that these were the octahedron and the icosahedron. Baez is supported by https://www.britannica.com/science/Platonic-solid and also by a comment in the German edition of Euclid's "Elements" (Ostwalds Klassiker der exakten Wissenschaften, vol. 235, p. 471) which relates this scholium to the Greek mathematician *Pappus* who lived around 300 AD at Alexandria.

[15] The convex hull of a subset $S \subset \mathbb{R}^n$ is the smallest convex set $C \subset \mathbb{R}^n$ with $S \subset C$.

on $\mathbb{R}^n$ which keeps the subset $P \subset \mathbb{R}^n$ invariant: $G_P = \{F \in E(n);\; F(P) = P\}$. A polytope is called *regular* if its symmetry group acts transitively on the set of flags. In particular, the set of vertices $v_1, ..., v_N$ and therefore also its mean value or center of gravity $s = \frac{1}{N}\sum_i v_i$ is invariant under every $F \in G_P$, which is, after all, an affine mapping (cf. Sect. 2.5). Because s can be chosen as the origin, G_P fixes the origin 0 and thus consists of linear isometries: $G_P \subset O(n)$. Each $(n-2)$-dimensional face S_2 lies in exactly one hyperplane H through the origin (namely the linear span of S_2), and the reflection at H belongs to G_P and interchanges the two $(n-1)$-dimensional faces adjacent to S_2.

If v is a vertex of P, then we denote the union of all faces of v adjacent to v as the *star* of v. This already determines the whole polytope uniquely because we generate all remaining faces by successive reflections along the already known $(n-2)$-dimensional faces.

The n-dimensional regular convex polytopes for $n = 2$ are the regular polygons, for $n = 3$ they are the classical Platonic solids. In each dimension $n \geq 3$ there are at least three regular convex polytopes:

1. The *simplex,* the generalization of the tetrahedron, with the $n + 1$ vertices $e_1, \ldots, e_{n+1}$ in the hyperplane $D := \{x \in \mathbb{R}^{n+1};\; \sum_i x_i = 1\} \subset \mathbb{R}^{n+1}$ which we can identify with $\mathbb{R}^n$,
2. the *cube* with the 2^n vertices $(\pm 1, \ldots, \pm 1) \in \mathbb{R}^n$,
3. the *co-cube*, the n-dimensional generalization of the octahedron, with the $2n$ vertices $\pm e_1, \ldots, \pm e_n$.

The corresponding polytope is in each case the convex hull of the set of vertices. Dodecahedron and icosahedron however do not belong to these three types. Are there such exceptions in other dimensions too? The answer comes as a surprise: Yes, but only in dimension $n = 4$.

The regular polytopes in $\mathbb{R}^4$ can be classified quite similarly to those in $\mathbb{R}^3$. They are bounded by congruent three-dimensional regular polytopes, Platonic solids. At each one-dimensional face ("edge") at least three such solids must meet, and as before the vertex angles, now the "edge angles" must add up to less than 360°, because the edge angles are equal to the vertex angles of a three-dimensional Platonic solid: the intersection of P with the perpendicular bisector on the edge (a hyperplane). For the tetrahedron the edge angles are a little more than 70° (Exercise 31); so we can put 3, 4, or 5 tetrahedra around an edge. For the cube the edge angles are 90°, and we can arrange only three of them around an edge (with four we would fill the three-dimensional space). For the octahedron the edge angle is also easy to calculate: One of the bounding polygons Σ_1 has the vertices e_1, e_2, e_3

and thus the normal vector[16] is $v_1 = e_1 + e_2 + e_3$. A neighboring face Σ_2 is spanned by the points $e_1, e_2, -e_3$ and therefore has the normal vector $v_2 = e_1 + e_2 - e_3$. The angle β between v_1 and v_2 satisfies $\cos\beta = \frac{\langle v_1, v_2\rangle}{|v_1||v_2|} = \frac{1}{3}$ so β is a little more than 70°.

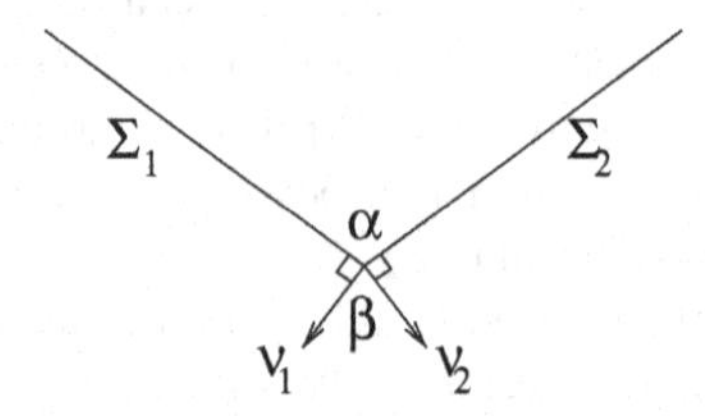

The edge angle we are looking for, $\alpha = 180° - \beta$, is therefore scarce 110°, and so 3 octahedra still fit around one edge. Also the edge angle of the dodecahedron is smaller than 120° and so we can still fit three dodecahedra around one edge. But for the icosahedron the edge angle is too large, $> 120°$. So we have found six possible arrangements: Around an edge there can be 3, 4, 5 tetrahedra, 3 cubes, 3 octahedra or 3 dodecahedra respectively.

The corresponding four-dimensional Platonic solids with tetrahedra as faces[17] are the simplex, the co-cube and a new polytope with 120 vertices, 720 edges, 1200 regular triangles and 600 tetrahedra, the *600-cell*. The three other possibilities with cubes, octahedra and dodecahedra as faces result in the four-dimensional cube and two new solids, the *24-cell* with 24 octahedra, 96 triangles, 96 edges and 24 vertices, as well as the *120-cell,* the *dual* is the 600-cell (the vertices of the *dual polytope* are the face centers of the given one, see Footnote 19), with 120 dodecahedra, 720 regular pentagons, 1200 edges and 600 vertices. The 24-cell (like the simplex) is *self-dual*, and its 24 vertices are the cube corners $(\pm 1, \pm 1, \pm 1, \pm 1)$ together with the co-cube corners $\pm 2e_i$, $i = 1, 2, 3, 4$ (cf. Exercise 33). The 600-cell also has these vertices and 96 more, which arise from the vector $(\pm\phi, \pm 1, \pm\phi^{-1}, 0)$ through all even permutations of the coordinates, where $\phi = \frac{1+\sqrt{5}}{2}$ is the golden ratio (cf. Berger [1], vol. II, pp. 32–36).

The fact that there are only simplices, cubes, and co-cubes in dimension five (and higher) can be seen again from the angles between the face normals of these six solids, cf. Exercise 31.

16 A *normal vector* ν on an affine plane in $\mathbb{R}^3$ is a vector perpendicular to the directions in the plane. For example, if the plane passes through three (affinely independent) points u, v, w, then ν is perpendicular to $v - u$ and $w - u$. The normal vector is determined only up to a real scalar; if one demands $|\nu| = 1$, there are still two possibilities.

17 The union of all faces adjacent to an edge, the *star* of the edge, determines the regular polytope as well as the star of a corner.

4.6 Symmetry Groups of Platonic Solids

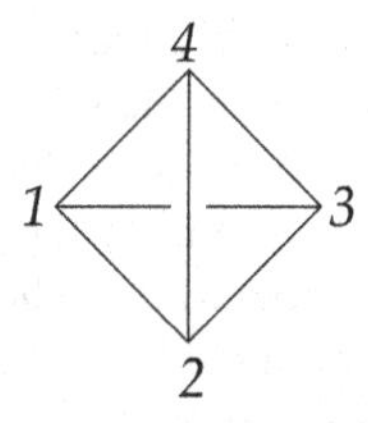

Which symmetries do the Platonic solids have? A *symmetry of a solid* is an isometry of the surrounding space, which leaves the solid (as a subset of this space) invariant. The best known symmetries are the reflections; the corresponding mirrors (fixed planes) are called *planes of symmetry.* At the tetrahedron (see figure) we immediately see the planes of symmetry; for example, the reflection along the plane containing the line 24 and the midpoint of the line segment $\overline{13}$ is a symmetry of the tetrahedron, which leaves the points 2 and 4 fixed and 1 and 3 interchanged. We can do the same with any other pair of points $i, j \in \{1, 2, 3, 4\}$. So the symmetry group of the tetrahedron contains all transpositions of two vertices, while leaving the other two vertices fixed. Since all permutations are compositions of transpositions (interchanges), the symmetry group of the tetrahedron contains all permutations of the set $\{1, 2, 3, 4\}$ and hence the whole group S_4 *(symmetric group).* On the other hand, each symmetry defines a permutation of the four vertices, and this permutation uniquely determines the symmetry, thus S_4 *is* the symmetry group of the tetrahedron. Similarly, one sees that S_n is the symmetry group of the n-dimensional simplex. If instead of all the symmetries we want to determine only the *rotations*, the symmetries which preserve the orientation (determinant 1), then we must restrict ourselves to those permutations of the set of vertices $\{1, 2, 3, 4\}$ which are compositions of an *even* number of transpositions, because the transpositions are plane reflections, so they have determinant -1. These permutations form the subgroup A_4, the *alternating group*; this is the *rotation group* of the tetrahedron.[18]

The next two Platonic solids, cube and octahedron, are closely related: The octahedron is obtained as the convex hull of the centers of the faces of the cube, and conversely the cube is obtained as the convex hull of the centers of the faces of the octahedron; the two solids are *dual* to each other, as one says. The number triples

(number of faces, number of edges, number of vertices)

are reversed by duality: (6,12,8) for the cube, (8,12,6) for the octahedron. Similarly, the dodecahedron and icosahedron are dual to each other with the triples (12,30,20) and (20,30,12), while the tetrahedron with triple (4,6,4) is dual to itself.[19] The

[18] The *alternating group* A_n is defined as the subgroup of the permutation group (*symmetric group*) S_n which is generated by the compositions of any two transpositions.

symmetry group of dual solids is the same, for a symmetry of the solid also preserves the set of face centers.

Let us first determine only the rotation group of the cube. There are as many rotations as positions of the cube with all edges parallel to the coordinate axes: The rotation group acts *simply transitively* on the set of positions. How many positions are there? We can put each of the 6 faces of the cube up, and then put every of the 4 edges of a face in front of us; this determines the position. So there are $6 \cdot 4 = 24$ positions and as many rotations of the cube. At each rotation the four spatial diagonals of the cube are permuted (see subsequent figure). There are $4! = 24$ such permutations and as many rotations, therefore the rotation group of the cube is exactly the full permutation group of the four space diagonals. For example, the transposition (23), the permutation of diagonals 2 and 3 without changing diagonals 1 and 4, corresponds to the 180-degree rotation in plane 14 spanned by the diagonals 1 and 4.

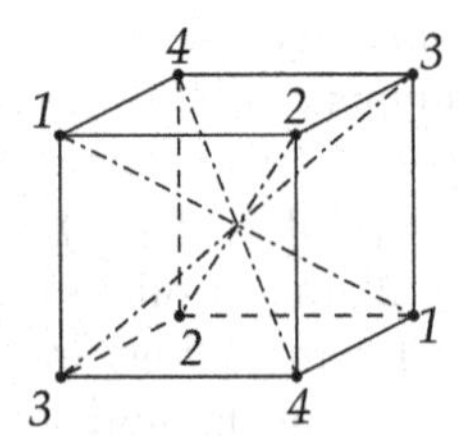

The full symmetry group is now easy to determine, because unlike the tetrahedron, the cube is invariant under the point reflection at the midpoint, the *antipodal map*, given by the matrix $-I = \begin{pmatrix} -1 & & \\ & -1 & \\ & & -1 \end{pmatrix}$. This is a "reflection" (det = −1), because $\det(-I) = (-1)^3 = -1$. Since the rotation group is a subgroup of index 2 in the symmetry group (the composition of every two reflections is a rotation), we have for the cube:

$$\text{symmetry group} = \text{rotation group} \times \{\pm I\} = S_4 \times \{\pm I\}.$$

The same argument applies to the icosahedron, which is, after all, also invariant under the antipodal map; thus we also have "symmetry group = rotation group $\times\{\pm I\}$". So only the rotation group is to be determined. However, the idea with the diagonals doesn't work so well this time: The icosahedron has 12 vertices and thus 6 diagonals, which would allow $6! = 720$ permutations. On the other hand there are only $20 \cdot 3 = 60$ positions and as many rotations of the icosahedron: Each of the 20 faces can be on top, and each of the three edges in front, then the position

[19] The concept of duality works the same way in higher dimensions. In dimension 4, the Platonic solids with three tetrahedra and with three octahedra per edge (simplex and 24-cell) are dual to themselves, the four-dimensional cube (3 cubes per edge) is dual to the four-dimensional octahedron (4 tetrahedra per edge), and the two remaining solids, the 600-cell and the 120-cell (5 tetrahedra and 3 dodecahedra per edge, respectively) are also dual to each other; cf. Hilbert and Cohn-Vossen [2], p. 144.

is determined. So the rotation group of the icosahedron is only an (interesting!) *subgroup* of the permutation group S_6. But there are other structures within the icosahedron that are permuted during rotations. We can position the icosahedron so that three pairs of edges are parallel to the coordinate axes ("paraxial").

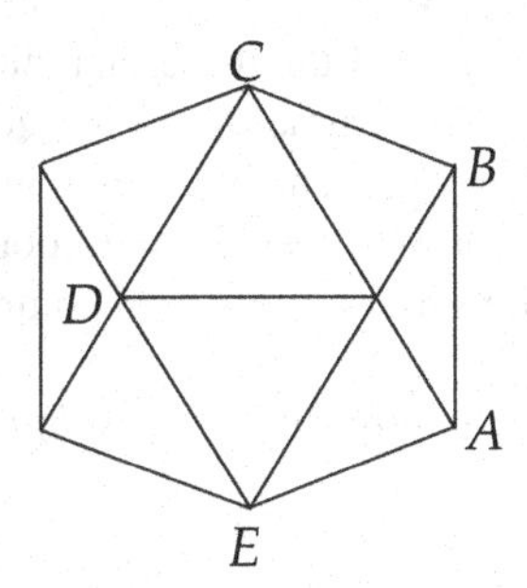

The figure shows the orthogonal projection of such a position onto a coordinate plane: The two vertical edges lie in the plane of projection, the front and back horizontal edges are parallel to this plane and coincide under projection, and the points C and E are projections of the edges perpendicular to the projection plane.[20] We can think of the icosahedron as including an octahedron whose 6 vertices lie exactly on the midpoints of these 6 edges.

After a 72° rotation of the icosahedron around one of its diagonals, 6 new edges come in paraxial position, and none of the previous edges remains in such position. We can therefore divide the 30 edges into 5 packets of 6 edges each, which are parallel or perpendicular to each other. In other words, there are 5 different embeddings of the octahedron into the icosahedron, each differing by a rotation. These 5 octahedra or 5 sets of edges are our objects that are permuted. Five objects admit 120 permutations; thus the rotation group of the icosahedron is a subgroup of S_5 of half order 60. There is only one such subgroup: the alternating group A_5. So the rotation group of the icosahedron is A_5 and the full symmetry group is $A_5 \times \{\pm I\}$. It has 120 elements like the S_5, but it is not S_5 because there is no odd permutation that commutes with all even permutations.

[20] The width and height of the figure are equal and in proportion of the golden ratio to the edge length. This is easy to see: The edge of the icosahedron is, after all, the edge of the pentagon which is formed by the five adjacent corners of a vertex, e.g. A, B, C, D, E, and the width and height of the figure (the distance between the parallel faces) is the diagonal in this pentagon. The edge AB and the diagonal CE lie in the plane of projection, so their lengths are preserved under projection. A similar projection exists for the dodecahedron, see Exercise 32.

4.7 Finite Subgroups of the Orthogonal Group, Patterns, and Crystals

The symmetry groups of Platonic solids ("Platonic groups") are finite subgroups of the full orthogonal group $O(3)$. Are there any other such groups, or is every finite subgroup of $O(3)$ contained in one of the three Platonic groups? There is certainly another series of such groups: the symmetries of a regular planar polygon (n-gon). This, too, can be understood as a (degenerated) Platonic solid, as a *dihedron*, an—albeit very flat—"solid" with only two faces (see Footnote 26). We want to show that this makes the list of finite subgroups of $O(3)$ complete:

Theorem 4.8 *Any finite subgroup of $O(3)$ is a subgroup of a Platonic group or dihedral group.*

Finite subgroups of $O(3)$ consist of finitely many isometric transformations of the unit sphere

$$\mathbb{S}^2 = \{x \in \mathbb{R}^3;\ |x| = 1\}.$$

Let us first consider an analogous problem for the plane $\mathbb{R}^2$ instead of the sphere $\mathbb{S}^2$. We might imagine a wallpaper pattern and its symmetry group G, a subgroup of the Euclidean group $E(2)$. Because of the infinite extension of the plane, the orbits of this group are no longer supposed to be finite (the wallpaper pattern continues periodically and repeats itself indefinitely often), but *discrete* instead.[21] We are interested in particular in the *rotations* in G (cf. Corollary 4.4), i.e., the elements $g \in G$ with an isolated fixed point A, the *center of rotation* for g. Actually, for any center of rotation A we consider the subgroup G_A of all rotations $g \in G$ with fixed point A. By discreteness G_A is finite (see Footnote 25). Thus there is some $g \in G_A$ with smallest rotation angle which must be $360°/n$ for some $n \in \mathbb{N}$.[22] This number n will be called the *order of the rotation center A*.

Note that fgf^{-1} is a rotation of the same order for every $f \in G$, and thus fA, the center of rotation for fgf^{-1}, has the same order as A. We therefore consider the orbit of a rotation center A of order n under the whole group G. If A is not fixed by G, its orbit consists of several points. We consider two points A and B in this orbit with minimal distance. There are rotations $g, h \in G$ about A, B, respectively, with rotation angle $360°/n$. We may assume that g rotates counter-clockwise and h rotates clockwise. By applying g to B and h to A we get two new rotation centers $A' = hA$ and $B' = gB$.

[21] A subset $D \subset \mathbb{R}^n$ is called *discrete* if for every $x \in D$ there exists an $\epsilon > 0$ with $D \cap B_\epsilon(x) = \{x\}$, where $B_\epsilon(x) := \{y \in \mathbb{R}^n;\ |x - y| < \epsilon\}$.

[22] Since $g^n = 1$ for some $n \in \mathbb{N}$, the rotation angle of g is $(k/n) \cdot 360°$ for relatively prime integers k, n. But $k = \pm 1$ since otherwise there were some integer j with $kj \equiv 1 \bmod n$ and g^j had smaller rotation angle $360°/n$, a contradiction to the choice of g.

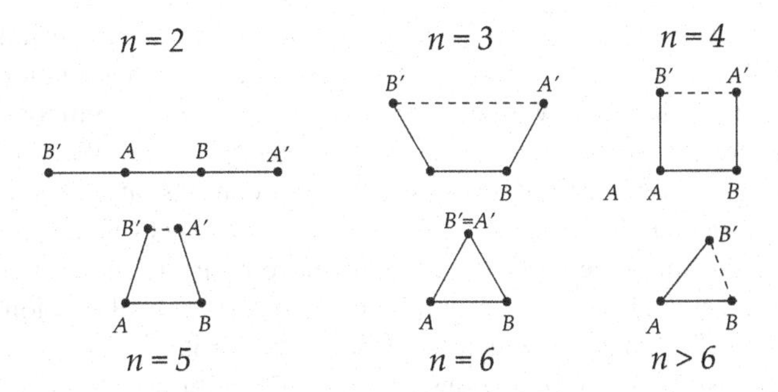

In the case $n = 5$ we see $|B' - A'| < |B - A|$, and in the case $n > 6$ we have $|B' - B| < |B - A|$; both properties are inconsistent with the assumption that A and B are centers of rotation of *minimal* distance. So these cases are impossible. On the other hand, the cases $n \in \{2, 3, 4, 6\}$ actually do occur:[23]

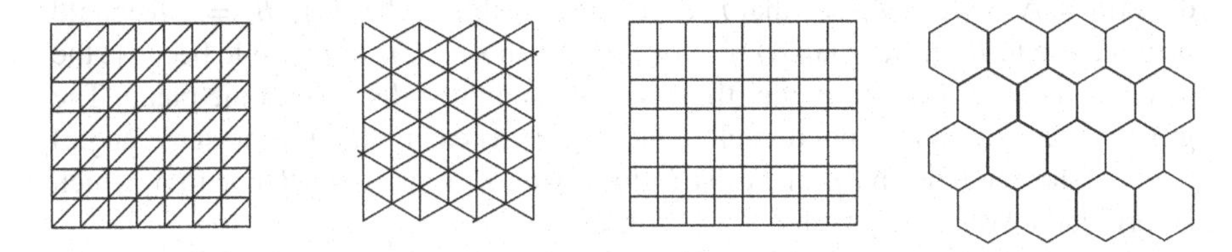

This result $n \in \{2, 3, 4, 6\}$ is called the *crystallographic restriction.* It is valid for all centers of rotation whose orbit consists of more than one point. In particular it holds if G contains a translation t_v because then every orbit is periodic in the v-direction and thus has even infinitely many elements.[24]

Now we replace the plane with the sphere $\mathbb{S}^2$ and apply the same ideas to a finite[25] subgroup G of the orthogonal group $O(3)$; this acts isometrically on $\mathbb{S}^2$ as

[23] Variations of these plane patterns have been used repeatedly for artistic purposes, especially for wallpaper, fabric, and tile designs. Especially the Islamic art , who was largely forbidden figurative representation, developed geometrical patterns and ornaments very widely, cf. Clévenot and Degeorge [13].

[24] Essentially, the argument also holds for discrete subgroups G of the *three*-dimensional Euclidean group $E(3)$, provided G contains a translation t_v. Here the centers of rotation are to be replaced by axes of rotation, and we must assume that the axis of rotation under consideration is not parallel to v. *Crystals* are triply periodic arrangements of atoms, hence their symmetry group contains three linearly independent translations, and we therefore find for each rotation axis a translation that is not parallel to its axis. Thus the crystallographic restriction holds also for the symmetry group of a crystal: Consider two parallel rotation axes with minimal distance (instead of two rotation centers with minimal distance); the rest of the argument remains unchanged. Thus, there are no crystals that admit a rotation of order 5 or ≥ 7, hence the name "crystallographic restriction". There are, however, crystal-like structures ("quasicrystals") that violate this restriction admitting fivefold symmetry, see, e.g., Eschenburg [12], Chap. 17. They, too, have a relation to Islamic art, see, e.g., Makovicky [15], as well as http://myweb.rz.uni-augsburg.de/~eschenbu/zula_saskiamayer.pdf.

[25] Since $O(n)$ is compact (bounded and closed in $\mathbb{R}^{n \times n}$), any discrete closed subset of $O(n)$ is finite, because infinitely many elements would contain a convergent subsequence.

the Euclidean group $E(2)$ does on the plane $\mathbb{R}^2$. As before, we can examine the G-orbit of a rotation center A of order n. Its G-orbit consists of other rotation centers of G with the same order n. Now rotation centers occur in pairs of antipodes $\{A, -A\}$ because they are the intersections of the axes of rotation (cf. Corollary 4.5) with the unit sphere $\mathbb{S}^2$. If the G-orbit of a rotation center A consists only of A and perhaps $-A$, then its order can be *any* $n \in \mathbb{N}$. But then all elements of G keep the axis $\mathbb{R}A$ invariant and thus also the plane $A^\perp$; we therefore return to the two-dimensional case with fixed origin. Every such group G either consists only of rotations, in which case it is the cyclic group C_n which is generated by a rotation with rotation angle $360°/n$, or it also contains reflections or upside-down turns interchanging A and $-A$, and then G is the twofold extension of C_n called *dihedral group* D_n.[26]

If besides A there is another rotation center $B \neq \pm A$ in the G-orbit of A then we can proceed as for the crystallographic restriction and assume that A and B have smallest distance within the orbit. Again we consider rotations g and h with rotation centers A and B and minimal rotation angles $360°/n$, but rotating in opposite directions. As before we see that $n > 6$ is impossible: Otherwise $B' = gB$ would be too close to B (closer than A). But this time also $n = 6$ is impossible because the distance on the sphere is smaller than the corresponding distance in the plane: The great circle arcs on the sphere incline more towards each other (at the same angle) than the straight lines in the plane. Therefore again B' is closer to B than to A, which is forbidden. We are left with $n \in \{5, 4, 3, 2\}$.

We will distinguish groups G all of whose elements have order 2 from those which (also) contain elements of higher order and therefore have rotation centers of order $n \in \{3, 4, 5\}$. We first treat this latter case. If $n = 5$, then $B' = gB$ and $A' = hA$ are closer together than A and B, which implies $A' = B'$. Then the points A, B, A' form an equilateral spherical triangle, and at each vertex five such triangles join (because of $n = 5$). The whole orbit of A consists of vertices of such triangles. These must be neatly contiguous, because if two of them intersected, the minimal distance condition would be violated. So the sphere is paved with such triangles, and their vertices are the vertices of a regular icosahedron, see Sect. 4.5.[27] The group G preserves this icosahedron since its vertices form an orbit of G, so G is a subgroup of the symmetry group of the icosahedron. The next case $n = 4$ is similar: Because the points B' and A' are closer together on the sphere than in the

[26] A *dihedron* ("two faces") is a polygonal disk which can be regarded as a degenerate Platonic solid with only two bordering polygons (front and back). The dihedral group is the group of all rotations and reflections of a regular plane polygon; by embedding the polygon in 3-space $\mathbb{R}^3$, we can think of the dihedral group as a subgroup of $SO(3)$, replacing the reflections by upside-down turns.

[27] The convex hull of the vertices is a solid bounded by equilateral triangles, with 5 triangles meeting at each vertex; this is the icosahedron, the 20-face solid. You can get the number 20 directly from Exercise 44: Since 5 congruent triangles meet at every vertex, each interior angle there is $2\pi/5$, so the angle sum of each spherical triangle is $6\pi/5$. By Exercise 44, then, the area of the spherical triangle is $F = 6\pi/5 - \pi = \pi/5$, that is, one-twentieth of the total area 4π of the sphere. Analogously $n = 4$ with $6\pi/4 - \pi = \pi/2 = 4\pi/8$ (octahedron) etc.

plane, they have smaller distances than A and B and therefore must be equal. Again the points A, B, A' form an equilateral spherical triangle, but this time only four triangles come together at each vertex, and we get the octahedron.

Now we assume $n = 3$. Then A, B, A', B' belong to an equilateral spherical polygon with oriented angles 120° which is contained in the G-orbit of A. This polygon must be simply closed since otherwise the minimal distance condition would be violated. Hence it is a closed regular spherical k-gon. But $k = 6$ (and a fortiori $k > 6$) is impossible: In the plane these points would form a regular hexagon with angles 120°, but on the sphere a regular k-gon has interior angles $> 120°$ for $k = 6$ and even larger for $k > 6$. (Exercise 44). So it is a spherical pentagon, quadrilateral, or triangle ($k = 5, 4, 3$), and at each vertex three of these polygons join; therefore the orbit of A consists of the vertices of either the dodecahedron, or the cube, or the tetrahedron, and G is contained in the corresponding platonic group.

Finally, if all elements of G are of order 2, then G must be commutative, because when $g, h \in G$, then $gh \in G$ has order 2 again and thus $gh = (gh)^{-1} = h^{-1}g^{-1} = hg$. Each $g \in G \setminus \{I\}$ is a reflection along a fixed space $F \subset \mathbb{R}^3$ which can have the dimensions $m = 0, 1, 2$. When $m = 0$, then $g = -I$ and thus g commutes with every $h \in O(3)$; we can therefore assume $-I \in G$ (otherwise we extend G by adding $-I$). When $m = 1$ then $-g = (-I) \circ g$ has a 2-dimensional fixed space, so it is a plane reflection. We can therefore assume that G is generated by $-I$ and some plane reflections which commute with each other. If g is a reflection along a plane E and h a reflection along another plane F, then g commutes with h if and only if g keeps the fixed space F of h invariant; therefore F must be perpendicular to E, i.e. the normal vectors of E and F are perpendicular to each other. Any further plane reflection $k \in G$ must have a mirror plane which is perpendicular to both E and F, so there is at most one such $k \in G$. We can choose the coordinate system such that e_1, e_2, e_3 are the normal vectors of the three mirror planes. Then G consists of the eight diagonal matrices diag$(\pm 1, \pm 1, \pm 1)$ in $O(3)$. But this is a subgroup of the octahedral group. □

4.8 Metric Properties of Conic Sections

An *ellipse* with *principal axes* a and b ($a > b > 0$) is obtained from the unit circle by applying the matrix $\begin{pmatrix} a & \\ & b \end{pmatrix}$; in doing so, the two coordinates of each circle point are multiplied by the factors a and b, hence stretched or compressed. If we undo this stretching (or compressing), a point (x, y) of the ellipse becomes the point $(\frac{x}{a}, \frac{y}{b})$ of the circle, satisfying the equation

$$\frac{x^2}{a^2} + \frac{y^2}{b^2} = 1 \tag{4.13}$$

which is the equation of the ellipse. The *hyperbola* with the *principal axes a* and *b*, on the other hand, satisfies the analogous equation

$$\frac{x^2}{a^2} - \frac{y^2}{b^2} = 1 . \tag{4.14}$$

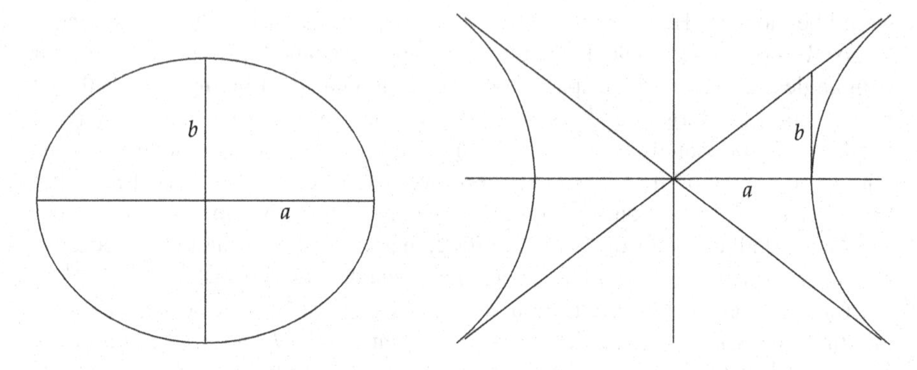

Moving these figures to the right so that the point $(a, 0)$ becomes their center we obtain instead the equations

$$\frac{(x-a)^2}{a^2} \pm \frac{y^2}{b^2} = 1 \tag{4.15}$$

(the points $(0, 0)$ and $(2a, 0)$ now satisfy these equations). If we now let both principal axes a, b go to ∞, but with the condition that always $b^2 = ac$ for some constant $c > 0$ (now also $a \leq b$ is allowed), then in the limit case one obtains the equation of the parabola

$$\pm y^2 = 2cx. \tag{4.16}$$

In fact, the Eq. (4.15) results in

$$\frac{x^2}{a^2} - \frac{2x}{a} = \mp \frac{y^2}{b^2},$$

and after multiplication with b^2 we get $\pm y^2 = 2cx - x^2 \cdot c/a$. Since $c/a \to 0$, it follows (4.16).

An ellipse has two so-called *focal points* F, F', which lie at a distance $f = \sqrt{a^2 - b^2}$ from the center on the long principal axis (see Exercise 35). They have the following property:

Theorem 4.9 *The sum of the distances from each point P on the ellipse to the focal points F, F′ is constant.*

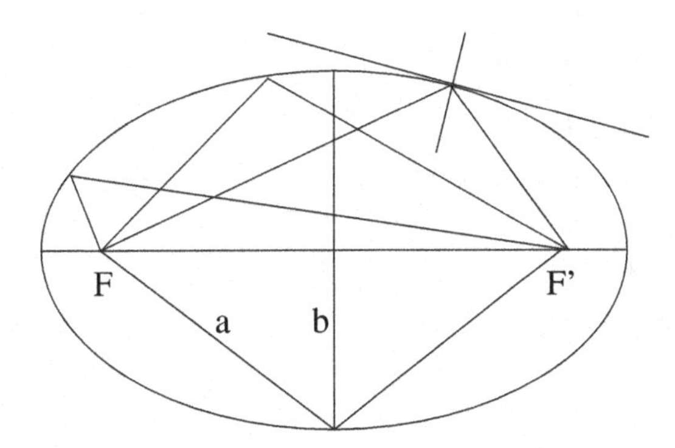

As a corollary we see that the angles between the tangent of the ellipse at P and the two lines PF and PF' are equal: If they were unequal, the reflection S at the tangent line would show that the sum of the distances on one side would be larger than on the other side, see figure (the broken path is longer).[28]

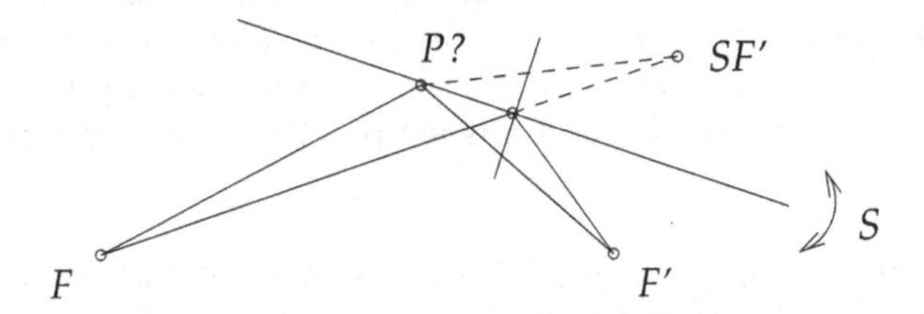

Light or sound waves that emanate from a focal point F and are reflected at the ellipse will therefore meet again at F'.

This property of Theorem 4.9 (as well as the corresponding properties of the parabola and the hyperbola) can be calculated from the defining equation $\frac{x^2}{a^2} + \frac{y^2}{b^2} = 1$, but in geometry it is more fun to read such a property from a figure. This is successful when we regard the ellipse (or parabola, or hyperbola) really as a *conic section*, as the intersection of a cone with a plane E in 3-space. It is somewhat easier for the imagination if we first replace the cone by a circular cylinder. To do this, we consider two spheres that just fit into the cylinder or cone, and that touch the plane from above and from below (*Dandelin's spheres*).[29]

Dandelin's Theorem: *The focal points F, F' of an ellipse considered as conic section are just the points of contact of the two Dandelin spheres with the plane E containing the ellipse.*

[28] We have thereby identified the ellipse with its tangent near the point of contact. The error is of 2nd order and therefore negligible.

[29] Germinal Pierre Dandelin, 1794 (Le Bourget)–1847 (Ixelles).

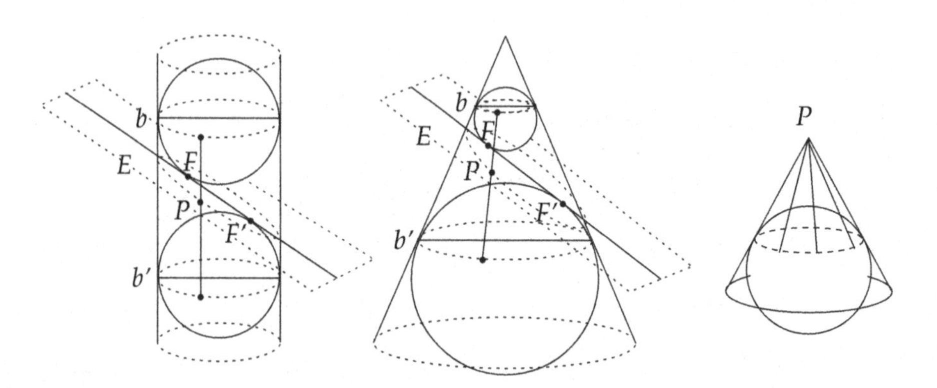

Proof Let us consider an arbitrary point P on the ellipse which is the intersection of the plane E with the surface of the cylinder or cone. The distance from P to F is the same as the distance from P to the circle of contact b on the upper sphere, because all tangent segments from P to the sphere have equal length (they form the generatrices of a circular cone, see figure on the right), and likewise the distance from P to F' equals the distance from P to the circle of contact b' on the lower sphere. The sum of the distances from P to F and F' is therefore equal to the distance of the two circles of contact, and this is independent of the ellipse point P, i.e., constant. □

Focal points with properties similar as in Theorem 4.9 exist also for the other conic sections as we see from the figures below.

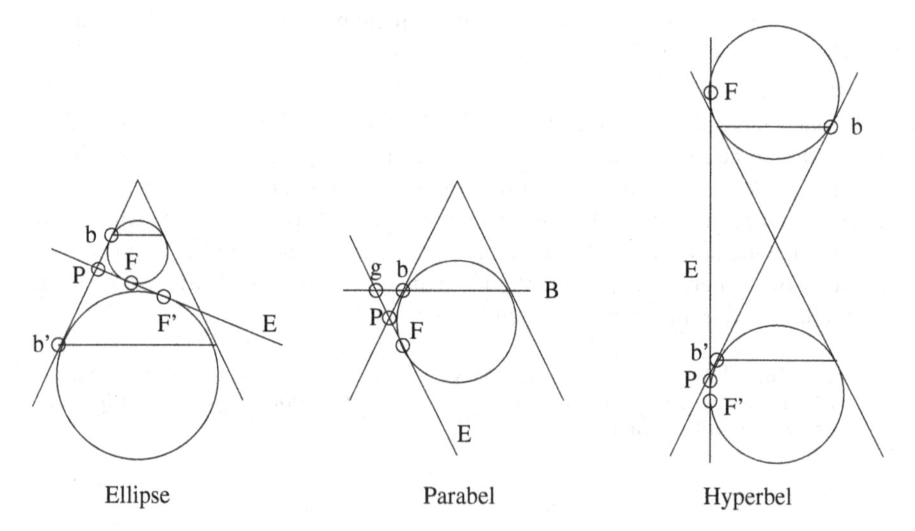

Ellipse Parabel Hyperbel

For the *hyperbola*, instead of the sum the *difference* of the distances to the focal points is constant: This constant is the "distance" of the two circles of contact, more precisely, the length of the segments of the generatrices on the double cone between the two circles of contact, see the right figure.

Also for a point P of the *parabola* (central figure above), the distance from P to F equals the distance to the circle of contact b. However, there is just one Dandelin sphere. But the circle b lies in a plane B meeting the plane E of the parabola in a straight line g called *directrix*, and because E is parallel to a generatrix of the cone, the distances from P to b (along a generatrix) and to g (along E) are equal (isosceles triangle Pgb). Thus the distances from P to the point F and to the line g are the same.[30]

If one considers a parallel g' to the directrix (see subsequent figure), the sum of the distances of P to F and to g' is equal to the constant distance of g and g'. Again, therefore, the two angles of P with the parabolic tangent in P must be the same. All light rays emanating from F are thus reflected at the parabola parallel to the axis, and a parallel beam incident from this direction is focused into the focal point; hence the technical application of the paraboloid[31] in headlights as well as in parabolic mirrors and satellite dishes.

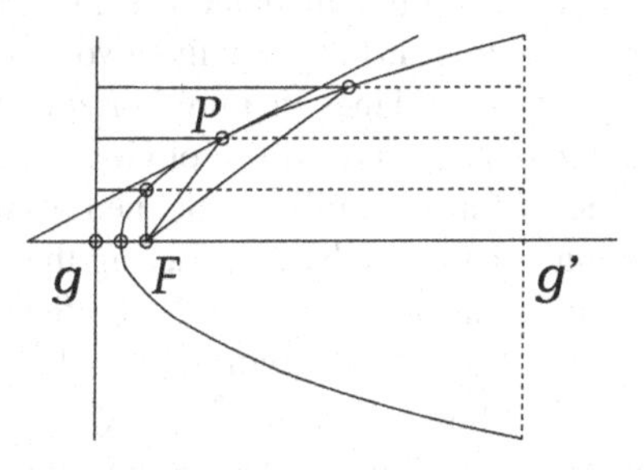

Due to these properties, it is possible to mechanically generate ellipse, parabola and hyperbola. The best known is the "gardener's construction" of the ellipse, in which a thread or rope is fixed at two points, the focal points, and pulled taut with the drawing instrument (left figure).

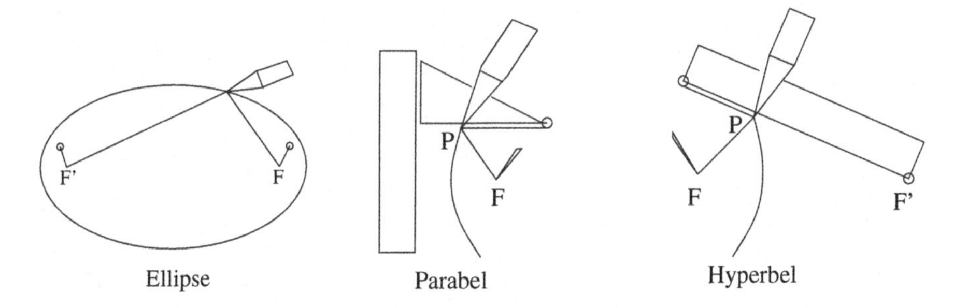

[30] A directrix also exists for ellipses and hyperbolas. As with the parabola, the directrix is the line of intersection of the plane E with the plane of the circle of contact for one of the spheres. But the distances from P to the focal point and to the directrix are no longer equal, but only proportional, because the slopes of the plane E and of the generators of the cone are no longer equal, however still independent of P; see Exercise 36.

[31] The *paraboloid* is the surface $\{(x, y, z) \in \mathbb{R}^3;\ z = x^2 + y^2\}$ which results from the rotation of a parabola about its axis.

For the parabola, the thread is fixed with one end at the focal point F and with the other end at a vertex of a right triangle (e.g. set square). Using the drawing instrument we pull the thread tight along the leg of the right triangle which is gliding with the other leg along the ruler (being the directrix g).

For the hyperbola one fixes one end of a ruler at a point F' on the drawing paper; the thread is attached to the other end of the ruler and to a second point F on the paper. Now pull it tight along the ruler while rotating the ruler about the point F'.

Remark Dandelin's construction is an example of how theorems on plane geometry are sometimes easier to prove using spatial (three-dimensional) geometry. Other examples were the theorems of Desargues and Brianchon (Sects. 3.5 and 3.7). Analogously, theorems on spatial geometry can sometimes be shown by using four-dimensional geometry. An example is the surface of an annulus (doughnut), also called *torus of revolution*: It is the surface of revolution D (see Exercise 38) which is obtained by rotating a circle in the half-plane $\{(x, z) \in \mathbb{R}^2;\ x > 0\}$ about the z-axis in xyz-space. Obviously this surface contains two sets of (plane) circles, the intersections of the doughnut with planes that are horizontal (perpendicular to the z-axis) or vertical (containing the z-axis). Not so obvious is a third family of circles on D which is skew (in fact, a fourth family is obtained from it by reflection). This can be seen almost without calculation by considering the image of the doughnut D under stereographic projection $\Phi : \mathbb{R}^3 \to \mathbb{S}^3$ (see Sect. 6.4). This image is a surface $T = \mathbb{S}^1_a \times \mathbb{S}^1_b \subset \mathbb{C} \times \mathbb{C} = \mathbb{R}^4$ (with $\mathbb{S}^1_a := \{w \in \mathbb{C};\ |w| = a\}$) with $a, b > 0$ and $a^2 + b^2 = 1$; thus T lies in $\mathbb{S}^3 \subset \mathbb{R}^4$. At T, besides $\mathbb{S}^1_a$ and $\mathbb{S}^1_b$, there is obviously another set of plane circles passing through every given point $(w_1, w_2) \in T \subset \mathbb{C}^2$: the *Hopf circles* $\mathbb{S}^1(w_1, w_2) = \{(\lambda w_1, \lambda w_2);\ \lambda \in \mathbb{S}^1 \subset \mathbb{C}\}$. Since the stereographic projection maps circles onto circles, the image of each Hopf circle under Φ^{-1} is a (skew) circle on the doughnut D. (https://www.youtube.com/watch?v=PYR9worLEGo)

5 Curvature: Differential Geometry

Abstract

Differential geometry deals with objects that are no longer "straight", such as curved lines and surfaces. The *curvature*, which measures the deviation from a straight line or a plane, is the central concept. While the curvature of a curve is given by a single number at each point, a surface (or hypersurface) requires a symmetric matrix whose eigenvalues are the "principal curvatures" of the surface; the eigenvectors are called principal curvature directions. We will unfold only a small part of this geometry, and only with a view to the following chapter, in which the simplest curved surfaces, the spheres, will play a central role. These will be characterized among all curved surfaces by the property that *all* tangential directions are principal curvature directions. For this we will study a class of curvilinear coordinate systems in space preserved by the angle-preserving (isogonal) mappings of the following chapter, namely, those in which all coordinate surfaces intersect perpendicularly. The tangents of the intersecting lines are then principal curvature lines for both intersecting coordinate surfaces.

5.1 Smoothness

So far we have dealt mainly with straight lines, planes, subspaces, or rectilinearly bounded objects such as triangles and polytopes. But already conic sections and quadrics show that not all geometrically interesting objects are rectilinear. *Differential geometry* (e.g. cf. [20]) deals with such curved objects that are no longer straight. However on a small scale they still look approximately like affine subspaces: If you consider just a short piece of a circle or a parabola, you might think it is a straight line, and until around 250 BC people thought that Earth was a flat

J.-H. Eschenburg, *Geometry – Intuition and Concepts*,
https://doi.org/10.1007/978-3-658-38640-5_5

disk.[1] The colloquial word for this property of having no irregularities on the small scale is "smooth", as opposed to "rough". The mathematical expression for this property is approximability by straight (linear) objects, and linear approximability is precisely *differentiability*.

How can we mathematically describe smooth objects such as curved lines or surfaces? Let us first consider how to describe *linear* objects, e.g. an m-dimensional linear subspace $W \subset \mathbb{R}^n$. There are basically only two kinds of description: either as *preimage* (*kernel*) or as *image*. The first description is as the solution set of a linear equation: $W = F^{-1}(0) = \{x \in \mathbb{R}^n;\ F(x) = 0\}$ for a surjective linear map $F : \mathbb{R}^n \to \mathbb{R}^{n-m}$. The second description is by *parametrization*, that is $W = \varphi(\mathbb{R}^m) = \{\varphi(u) : u \in \mathbb{R}^m\}$ for some injective linear map $\varphi : \mathbb{R}^m \to \mathbb{R}^n$. In the latter case, the points of W are "numbered" or, as one says, *parametrized* by the elements of $\mathbb{R}^m$.

Also our smooth objects can be described in these two ways; we only have to replace the word "linear" by "differentiable". The second description by parametrization is more explicit, therefore it is preferred in the present book[2] as in [21].

We recall that a mapping $\varphi : \mathbb{R}^m \to \mathbb{R}^n$ is called *differentiable* if it can be approximated at each point $u \in \mathbb{R}^m$ by a linear mapping $L : \mathbb{R}^m \to \mathbb{R}^n$ in the following sense: For all $h \in \mathbb{R}^m$ we have

$$\varphi(u+h) - \varphi(u) = Lh + o(h), \quad o(h)/|h| \xrightarrow{h \to 0} 0. \tag{5.1}$$

The linear mapping L depends on u and is called *derivative* or *differential* or *Jacobian matrix*[3] of φ at the point u; instead of L we usually write $d\varphi_u$ (and in the case $m = 1$ also $\varphi'(u)$):

$$L = d\varphi_u. \tag{5.2}$$

The columns of this matrix are the *partial derivatives*, which we denote $\frac{\partial \varphi(u)}{\partial u_i}$ or shorter $\partial_i \varphi(u)$ or still shorter $\varphi_i(u)$ $(i = 1, \dots, m)$:

$$\varphi_i(u) = \partial_i \varphi(u) = d\varphi_u e_i \in \mathbb{R}^n. \tag{5.3}$$

It is rare that a differentiable mapping φ is defined on the whole space $\mathbb{R}^m$; often the domain of definition is a connected *open subset* in $\mathbb{R}^m$, i.e. a subset $\mathbb{R}^m_o \subset \mathbb{R}^m$ which for each of its points contains also a small ball around that point, and any two points in $\mathbb{R}^m_o$ can be connected by a curve contained in $\mathbb{R}^m_o$ (*connectivity*). We will

[1] Eratosthenes of Cyrene, 276–194 BC, conjectured the spherical shape of Earth and even determined its circumference, see Exercise 37.

[2] An exception is Exercise 39. In the solution to this exercise in Chap. 9 we also describe the transition between the two descriptions, using the inverse function theorem.

[3] Carl Gustav Jacob Jacobi, 1804 (Potsdam)–1851 (Berlin).

also assume that the partial derivatives φ_i themselves are again continuous or even differentiable with continuous derivative (*continuously differentiable*). In the latter case we can also consider partial derivatives of φ_i which we call φ_{ij}; it holds that

$$\varphi_{ij} = \varphi_{ji}. \tag{5.4}$$

A differentiable mapping $\varphi : \mathbb{R}^m_o \to \mathbb{R}^n$ is called an *immersion* if for each $u \in \mathbb{R}^m_o$ the linear map $d\varphi_u : \mathbb{R}^m \to \mathbb{R}^n$ is injective, that is, if the partial derivatives $\varphi_1(u), \dots, \varphi_m(u)$ (the columns of $d\varphi_u$) at each point u are linearly independent vectors in $\mathbb{R}^n$. Because of (5.1), an immersion near u almost looks like an injective linear map (except for an additive constant),[4] so its image near $\varphi(u)$ is almost an affine subspace. These are the *smooth objects* of differential geometry: images of immersions. However, in geometry the actual parametrization φ is not important; it is, after all, only the naming of the points of the interesting object: the image of φ. As in the linear case, there are many other parametrizations describing the same object, viz. $\tilde{\varphi} = \varphi \circ \alpha$ for a *diffeomorphism* (an invertibly differentiable map) $\alpha : \mathbb{R}^m_1 \to \mathbb{R}^m_o$, defined on another connected open domain $\mathbb{R}^m_1 \subset \mathbb{R}^m$. Such a mapping α renames only the points of the image of φ; we call it *parameter change*. All geometric statements will be invariant under parameter changes. An immersion with dimension $m = 1$ will be called a *curve*, with $m = 2$ a *surface* and with $m = n - 1$ a *hypersurface*.

The *tangent space* of an immersion $\varphi : \mathbb{R}^m_o \to \mathbb{R}^n$ at a parameter point $u \in \mathbb{R}^m_o$ is the linear subspace $T_u := \operatorname{im} d\varphi(u) \subset \mathbb{R}^n$ with the basis $\varphi_1(u), \dots, \varphi_m(u)$. The orthogonal complement $N_u := (T_u)^\perp$ is called the *normal space* of φ in u. Tangent space and (consequently) normal space are independent of the choice of parametrization, because for $\tilde{\varphi} = \varphi \circ \alpha$ and $u = \alpha(\tilde{u})$ the chain rule gives

$$d\tilde{\varphi}_{\tilde{u}} = d\varphi_u d\alpha_{\tilde{u}}, \tag{5.5}$$

and thus $\operatorname{im} d\tilde{\varphi}_{\tilde{u}} = \operatorname{im} d\varphi_u$ because $d\alpha_{\tilde{u}}$ is invertible. If $\operatorname{im} \varphi$ is not "straight" or "plane", i.e. part of an m-dimensional subspace of $\mathbb{R}^n$, then T_u and N_u depend on u; their first derivatives as functions of u will be described in the next section as *curvatures*.

5.2 Fundamental Forms and Curvatures

For the sake of simplicity we restrict our attention to hypersurfaces ($n = m + 1$). Then the normal space is one-dimensional, i.e. generated by only one vector (*normal vector*), and there is a differentiable mapping $\nu : \mathbb{R}^m_o \to \mathbb{R}^n$ with $N_u = \mathbb{R} \cdot \nu(u)$ for

[4] The variable in (5.1) is h, while u and $\varphi(u)$ are constants.

all $u \in \mathbb{R}^m_o$.[5] Without loss of generality we may assume additionally $|\nu(u)| = 1$ for all u (if necessary we have to pass to $\nu/|\nu|$); such a map ν is called *unit normal field* or *unit normal*, also called *Gauss map*.[6]

For all $i, j \in \{1, \dots, m\}$ we consider the following functions on $\mathbb{R}^m_o$,

$$\begin{aligned} g_{ij} &= \langle \varphi_i, \varphi_j \rangle = & &\langle d\varphi.e_i, d\varphi.e_j \rangle, \\ h_{ij} &= \langle \varphi_{ij}, \nu \rangle \overset{*}{=} -\langle \varphi_i, \nu_j \rangle &= -&\langle d\varphi.e_i, d\nu.e_j \rangle, \end{aligned} \tag{5.6}$$

at "$\overset{*}{=}$" note $\langle \varphi_i, \nu \rangle = 0$ and therefore $\langle \varphi_{ij}, \nu \rangle + \langle \varphi_i, \nu_j \rangle = \partial_j \langle \varphi_i, \nu \rangle = 0$. For each $u \in \mathbb{R}^m_o$, the matrices $(g_{ij}(u))$, $(h_{ij}(u))$ belong to symmetric bilinear forms $g(u)$, $h(u)$ on $\mathbb{R}^m$:

$$\begin{aligned} g(u)(v, w) &= \sum g_{ij}(u) v_i w_j &= &\langle d\varphi_u v, d\varphi_u w \rangle, \\ h(u)(v, w) &= \sum h_{ij}(u) v_i w_j &= -&\langle d\varphi_u v, d\nu_u w \rangle. \end{aligned} \tag{5.7}$$

We call g the *first fundamental form* and h the *second fundamental form* of the immersion φ. The first fundamental form is nothing but the scalar product on the tangent space $T_u \subset \mathbb{R}^{m+1}$ where T_u is identified with $\mathbb{R}^m$ by the basis $\varphi_1(u), \dots, \varphi_m(u)$. It describes the distortion of length for the transition from the parameter domain $\mathbb{R}^m_o$ to the hypersurface $\operatorname{im} \varphi \subset \mathbb{R}^{m+1}$ by the parametrization φ. The second fundamental form describes the variation of the unit normal $\nu(u)$ and thus of the normal space N_u as a function of u.

In the simplest case $m = 1$ (plane curves) we have $g = g_{11} = |\varphi'|^2$ and $h = h_{11} = \langle \varphi'', \nu \rangle = -\langle \varphi', \nu' \rangle$. Of course, h depends on the parametrization: The larger the velocity $|\varphi'|$ by which the curve is traversed, the larger is $|h|$. A parameter-independent quantity is the quotient $\kappa = h/g$. In fact, if $\tilde\varphi = \varphi \circ \alpha$ is another parametrization, then $\tilde\nu = \nu \circ \alpha$. According to the chain rule we have $\tilde\varphi' = \varphi'\alpha'$ and $\tilde\nu' = \nu'\alpha'$ and thus $\tilde h = (\alpha')^2 h$ and $\tilde g = (\alpha')^2 g$, so $\tilde h / \tilde g = h/g$.[7] This quantity

$$\kappa = \frac{h}{g} = \frac{\langle \varphi'', \nu \rangle}{|\varphi'|^2} = \frac{-\langle \varphi', \nu' \rangle}{|\varphi'|^2} \tag{5.8}$$

[5] For every $u \in \mathbb{R}^{n-1}_o$ the mapping $x \mapsto \det(\varphi_1(u), \dots, \varphi_{n-1}(u), x)$ is a nonzero linear form on $\mathbb{R}^n$ (a row vector), differentiably depending on u. The corresponding column vector is $\nu(u)$, so $\det(\varphi_1(u), \dots, \varphi_{n-1}(u), x) = \langle \nu(u), x \rangle$. In the case $n = 3$ this is the *cross product*: $\nu = \varphi_1 \times \varphi_2$. Obviously, $\langle \nu, \varphi_i \rangle = \det(\varphi_1, \dots, \varphi_i, \dots, \varphi_{n-1}, \varphi_i) = 0$, so $\nu(u) \in N_u$.

[6] Johann Carl Friedrich Gauss, 1777 (Braunschweig)–1855 (Göttingen), wrote in 1828 in connection with the survey of the Kingdom of Hanover, which he directed, the work "Disquisitiones generales circa superficies curvas" (https://archive.org/details/disquisitionesg00gausgoog), English: "General investigations of curved surfaces" (https://www.gutenberg.org/files/36856/36856-pdf.pdf). There he introduces this map and the related Gaussian curvature. For a mathematical appreciation of this fundamental work, see M. Spivak: A Comprehensive Introduction to Differential Geometry, Vol. 2, Publish or Perish Inc., 1970, 1999.

[7] For convenience we have omitted the arguments: The functions φ', ν', g, h are to be taken at location u and $\tilde\varphi', \tilde\nu', \alpha', \tilde g, \tilde h$ at location $\tilde u$ with $\alpha(\tilde u) = u$.

is called the *curvature* of the plane curve $\varphi : \mathbb{R}^1_o \to \mathbb{R}^2$. A circle of radius r with the unit normal pointing inwards $\nu(u) = -\varphi(u)/r$ has the curvature $\kappa = 1/r$. In fact, $\varphi(u) = (r\cos u, r\sin u)$ and $\nu(u) = -(\cos u, \sin u)$, therefore $h(u) = -\langle\varphi'(u), \nu'(u)\rangle = r$ and $g(u) = |\varphi'(u)|^2 = r^2$. In general, $1/|\kappa|$ is the radius of the best approximating circle.[8]

In arbitrary dimension m we can do the same, because $g = (g_{ij})$ is a positive definite symmetric matrix and therefore invertible. Analogously to (5.8) we define the following matrix A, called *Weingarten map*,[9] which generalizes the curvature of a curve:

$$A := g^{-1}h\,. \tag{5.9}$$

Thus $gA = h$ and hence

$$g(Av, w) = h(v, w) \tag{5.10}$$

for all $v, w \in \mathbb{R}^m$ (the dependence on $u \in \mathbb{R}^m_o$ is suppressed in the notation). Because of the symmetry $h(v, w) = h(w, v)$, the matrix A is self-adjoint with respect to the inner product given by g. Thus this matrix is real diagonalizable with a g-orthonormal eigenbasis; the eigenvalues $\kappa_1, \ldots, \kappa_m$ are called *principal curvatures* and their eigenvectors in $\mathbb{R}^m$ (sometimes also in $T_u = d\varphi_u(\mathbb{R}^m)$) are called *principal curvature directions*. After a parameter change $\tilde\varphi = \varphi \circ \alpha$, the Weingarten map A_u is conjugated by $d\alpha_{\tilde u}$, which does not change the eigenvalues (principal curvatures).

---

[8] The circles touching a plane curve φ at the point $x = \varphi(u)$ (i.e. pass through x and have there the same tangent as the curve) are separated into two families: those which are to the right and those which are to the left of the curve near the point x. There is precisely one circle separating these two families: This is the circle of curvature (cf. Hilbert and Cohn-Vossen [2]).

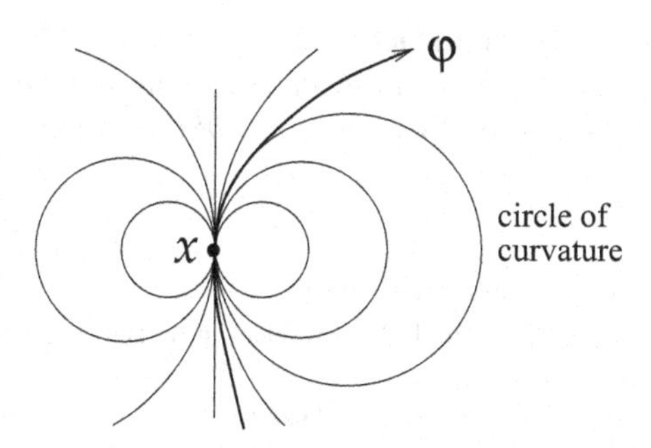

[9] Julius Weingarten, 1836 (Berlin)–1910 (Freiburg/Br.).

Even more important than the principal curvatures are their symmetric functions, in particular the arithmetic mean and the product:

$$H = \frac{1}{m}\sum \kappa_i = \frac{1}{m}\text{trace}\, A, \quad K = \prod \kappa_i = \det A, \tag{5.11}$$

called *mean curvature* and *Gauss-Kronecker curvature*.[10] We can only indicate the geometric meaning of these quantities here. For surfaces ($m = 2$), H describes the change of the area under deformations of the surface; in particular, surfaces with $H = 0$ cannot be made smaller by local deformations; therefore they are called *minimal surfaces* . On the other hand, K describes the area of the image of ν compared to the area of the image of φ. The same applies to any dimension m. The most important contribution of Gauss to differential geometry was the proof that for surfaces K depends only on g and thus remains invariant under *bending* deformations, for which g does not change. For example, if one rolls a flat sheet of paper into the shape of a cylinder or cone, the Gaussian curvature remains constantly zero, because one principal curvature remains zero. This observation was actually the birth of a new branch of geometry, now called *Riemannian geometry*, in which only a variable scalar product $g(u)$ (called *Riemannian metric*) is given, depending differentiably on $u \in \mathbb{R}^m_o$ (more generally on an m-dimensional manifold). In fact it was *Bernhard Riemann*,[11] a student of Gauss, who took this step in 1854 in his famous inaugural lecture "Über die Hypothesen, welche der Geometrie zugrunde liegen".[12] Without this development, for example, the theory of General Relativity by *Albert Einstein*[13] would not have been conceivable.

5.3 Characterization of Spheres and Hyperplanes

An *umbilic point* of a hypersurface $\varphi : \mathbb{R}^m_o \to \mathbb{R}^{m+1}$ is a point $u \in \mathbb{R}^m_o$ (or its image $\varphi(u) \in \mathbb{R}^{m+1}$) where the Weingarten map A_u is a multiple of the identity, that is $A_u v = \kappa(u)v$ for all $v \in \mathbb{R}^n$. The hypersurface φ has the *umbilic point property* and is called *umbilic hypersurface* if *all* $u \in \mathbb{R}^m_o$ are umbilic points.

Theorem 5.1 (Umbilic Hypersurfaces) *Let $\varphi : \mathbb{R}^m_o \to \mathbb{R}^{m+1}$ for $m \geq 2$ be a C^3-hypersurface (all second partial derivatives φ_{ij} are continuously differentiable) with the umbilic point property. Then the image of φ is contained in a sphere or a hyperplane in $\mathbb{R}^{m+1}$.*

---

[10] Leopold Kronecker, 1823 (Liegnitz)–1891 (Berlin).

[11] Georg Friedrich Bernhard Riemann, 1826 (Breselenz near Dannenberg/Elbe)–1866 (Selasca near Verbania, Lago Maggiore).

[12] "On the Hypotheses which lie at the Bases of Geometry", see the German text and its English translation by William Kingdon Clifford at www.maths.tcd.ie/pub/HistMath/People/Riemann/Geom/.

[13] Albert Einstein, 1879 (Ulm)–1955 (Princeton).

Proof By our assumption we have $h_{ij} = \kappa\, g_{ij}$. Since $h_{ij} = -\langle \nu_i, \varphi_j \rangle$ and $g_{ij} = \langle \varphi_i, \varphi_j \rangle$, it follows $\langle \nu_i, \varphi_j \rangle = -\kappa \langle \varphi_i, \varphi_j \rangle$ and thus, since $\nu_i \perp \nu$ (because of $\langle \nu, \nu \rangle = 1$) are linear combinations of $\varphi_1, \dots, \varphi_m \perp \nu$,

$$\nu_i = -\kappa \varphi_i. \tag{5.12}$$

Differentiation of (5.12) results in

$$\nu_{ij} = -\left(\kappa_j \varphi_i + \kappa \varphi_{ij}\right),$$
$$\nu_{ji} = -\left(\kappa_i \varphi_j + \kappa \varphi_{ji}\right).$$

Since $\nu_{ij} = \nu_{ji}$ and $\varphi_{ij} = \varphi_{ji}$ (because φ is three times and ν still twice continuously differentiable), it follows that

$$\kappa_i \varphi_j = \kappa_j \varphi_i. \tag{5.13}$$

Since the partial derivatives φ_i and φ_j for $i \neq j$ (here we need the dimensional assumption $m \geq 2$) are linearly independent, the coefficients in (5.13) vanish. So all partial derivatives κ_i vanish, and therefore κ is a constant.

(From now on also $m = 1$ is allowed.) We distinguish the cases $\kappa = 0$ and $\kappa \neq 0$. In the case $\kappa = 0$ we get $\nu_i = 0$ from (5.12), so ν is a constant unit vector. For a fixed parameter point u_o we set $s = \langle \varphi(u_o), \nu \rangle$. Then the image of φ is contained in the hyperplane $H = \{x \in \mathbb{R}^{m+1};\ \langle x, \nu \rangle = s\}$. In fact, each parameter point $u \in \mathbb{R}^m_o$ can be connected to u_o by a differentiable curve $t \mapsto u(t)$, and $\frac{d}{dt}\langle \varphi(u(t)), \nu \rangle = \langle d\varphi_{u(t)} u'(t), \nu \rangle = 0$, because $d\varphi_{u(t)} u'(t) \in \operatorname{im} d\varphi_{u(t)} = T_{u(t)} \perp \nu$. So $\langle \varphi(u(t)), \nu \rangle = const = s$ and thus $\varphi(u) \in H$.

In case $\kappa \neq 0$ we may assume $\kappa > 0$ (otherwise we pass to the normal $-\nu$). We set $r = 1/\kappa$. By (5.12) we have $\nu_j = -\frac{1}{r}\varphi_j$ and thus $(\varphi + r\nu)_j = \varphi_j + r\nu_j = 0$. Therefore $\varphi + r\nu = const =: M$ and so $|\varphi - M| = |r\nu| = r$, hence $\operatorname{im} \varphi$ is contained in the sphere of radius r with center M. □

5.4 Orthogonal Hypersurface Systems

A C^2-map $\Phi : \mathbb{R}^n_o \to \mathbb{R}^n$ (same dimension) is called *orthogonal parametrization* if the partial derivatives Φ_i are everywhere nonzero and perpendicular to each other: $\Phi_i \perp \Phi_j \neq 0$ for $i \neq j$.

An example are the *spherical coordinates* $\Phi : (0, \infty) \times \mathbb{R} \times (0, \pi) \to \mathbb{R}^3$,

$$\Phi(r, t, \theta) = r(\sin\theta \cos t,\ \sin\theta \sin t,\ \cos\theta). \tag{5.14}$$

In geometric terms r is the distance of the point $x = \Phi(r, t, \theta)$ from the origin, t the angle between the x_1-axis and the projection of x into the x_1x_2-plane, and

θ is the angle between x and the x_3-axis. This can be extended to any dimension n: First, construct a parametrization of the unit sphere in the $\mathbb{R}^n$ with orthogonal partial derivatives, i.e. $\varphi : \mathbb{R}^{n-1}_o \to \mathbb{R}^n$ with $\operatorname{im}\varphi \subset \mathbb{S}^{n-1} = \{x \in \mathbb{R}^n;\ |x| = 1\}$ and $\varphi_i \perp \varphi_j$ for $i \neq j$. This is done by induction over the dimension n: For $n = 2$ one sets $\varphi(t) = (\cos t, \sin t)$, and if one has already found such a parametrization $\bar{\varphi} : \mathbb{R}^{n-2}_o \to \mathbb{S}^{n-2}$ then one defines $\varphi : \mathbb{R}^{n-1}_o := \mathbb{R}^{n-2}_o \times (0, \pi) \to \mathbb{S}^{n-1}$, $\varphi(t, \theta) = \bar{\varphi}(t)\sin\theta + e_n \cos\theta$ for $t = (t_1, \ldots, t_{n-2}) \in \mathbb{R}^{n-2}_o$. Now one sets $\Phi : (0, \infty) \times \mathbb{R}^{n-1}_o \to \mathbb{R}^n$, $\Phi(r, u) = r\varphi(u)$ for all $u \in \mathbb{R}^{n-1}_o$.

As the example shows, an orthogonal parametrization Φ is not always invertible: The map $t \mapsto (\cos t, \sin t) : \mathbb{R} \to \mathbb{S}^1$ is not, because it wraps the line $\mathbb{R}$ around the circle $\mathbb{S}^1$ infinitely often. But since at every point $w \in \mathbb{R}^n_o$ the derivative $d\Phi_w$ is invertible (the n partial derivatives, the columns of $d\Phi_w$, are orthogonal and not equal to 0, thus linearly independent), Φ is invertible in a perhaps smaller neighborhood $\mathbb{R}^n_1$ around w, according to the inverse function theorem. We will therefore assume in addition (without restriction of generality) that Φ is invertible.

One interprets an orthogonal parametrization $\Phi : \mathbb{R}^n_o \to \mathbb{R}^n$ also as *orthogonal hypersurface system*: If one restricts Φ to a coordinate hyperplane $\mathbb{R}^{n-1}_{i,s} := \{w \in \mathbb{R}^n_o;\ w_i = s\}$ with $i \in \{1, \ldots, n\}$ and $s \in \mathbb{R}$, one obtains a hypersurface $\varphi^{i,s}$ with normal vector Φ_i, so in total n families of hypersurfaces $\varphi^{1,s}$ to $\varphi^{n,s}$. Through each point passes exactly one hypersurface of every family, and their normal vectors are perpendicular to each other, since $\Phi_i \perp \Phi_j$ for $i \neq j$. In the example of spherical coordinates in $\mathbb{R}^3$, the three hypersurfaces are the concentric spheres $r = const$, the vertical half planes $t = const$ and the circular cones $\theta = const$. Another example, the family of *orthogonal quadrics,* is described in Exercise 39. Orthogonal hypersurface systems have great geometric importance, partly because one knows their principal curvature directions:

Theorem 5.2 *If $\Phi : \mathbb{R}^n_o \to \mathbb{R}^n$ is an orthogonal parametrization with associated orthogonal hypersurface families $\varphi^{i,s}$, the principal curvature directions of the hypersurfaces $\varphi^{i,s}$ are the canonical basis vectors e_j, $j \neq i$.*

Proof We fix i and s and consider the hypersurface $\varphi = \varphi^{i,s}$. Their unit normal is $\nu = \Phi_i/|\Phi_i|$; the fundamental forms are $g_{jk} = \langle \varphi_j, \varphi_k \rangle$ and $h_{jk} = -\langle \nu_j, \varphi_k \rangle$ with $j, k \neq i$. We want to show that e_j is an eigenvector of $g^{-1}h$, that is $g^{-1}he_j$ is a multiple of e_j, or he_j is a multiple of ge_j. This in turn means $h_{jk} = t_j g_{jk}$ for some $t_j \in \mathbb{R}$ and for all $k \neq i$. Thus we must show that ν_j is a multiple of φ_j. By orthogonality of the partial derivatives $\varphi_k = \Phi_k$ for $k \neq i, j$ this claim is equivalent to $\nu_j \perp \nu_k$ for all $k \neq i, j$. Now $\nu = \frac{1}{|\Phi_i|}\Phi_i$ and thus $\langle \nu_j, \Phi_k \rangle = (\frac{1}{|\Phi_i|})_j \langle \Phi_i, \Phi_k \rangle + \frac{1}{|\Phi_i|}\langle \Phi_{ij}, \Phi_k \rangle$. Because $\langle \Phi_i, \Phi_k \rangle = 0$, it remains to show $\langle \Phi_{ij}, \Phi_k \rangle = 0$ for any three different indices i, j, k.

This holds because the expression depending on three indices $S_{ijk} = \langle \Phi_{ij}, \Phi_k \rangle$ has "too many" symmetries: On the one hand, we can interchange i and j (transposition (12)) without changing the value of S_{ijk}, because $\Phi_{ij} = \Phi_{ji}$. On the other hand, when we interchange i and k (transposition (13)), the sign changes:

We have $\langle \Phi_i, \Phi_k \rangle = 0 = const$ which implies $S_{ijk} = \langle \Phi_{ij}, \Phi_k \rangle = -\langle \Phi_i, \Phi_{kj} \rangle = -S_{kji}$. But the symmetry and the "antisymmetry" do not get along with each other. After all, what happens to S_{ijk} when we interchange the missing index pair j,k (transposition (23))? We can compute this in two ways: $(23) = (12)(13)(12)$ and $(23) = (13)(12)(13)$ *("braid relation")*, and hence:

$$S_{ijk} \overset{(12)}{=} S_{jik} \overset{(13)}{=} -S_{kij} \overset{(12)}{=} -S_{ikj},$$

$$S_{ijk} \overset{(13)}{=} -S_{kji} \overset{(12)}{=} -S_{jki} \overset{(13)}{=} S_{ikj}.$$

Thus $S_{ikj} = -S_{ikj}$, and thus the expression is zero for all distinct i, k, j. □

Angle: Conformal Geometry

6

Abstract

Distances can also be used to measure angles; for example, a triangle with side lengths 3, 4, 5 is right-angled (why?), which was used already by the ancient Egyptians in order to construct right angles. Conversely, however, distances cannot be determined only from angles. But there is a geometry with the angle as its only basic notion, called *conformal* geometry; it is much less known than metric geometry. Its "isomorphisms" are the conformal (i.e. angle-preserving) mappings. A great surprise is Liouville's Theorem: In dimension 2 there is an infinite-dimensional family of conformal mappings, namely all holomorphic and antiholomorphic complex functions. But in dimension 3 (and higher), conformal mappings automatically preserve the set of spheres and planes in space; such mappings form a family with finitely many parameters and can easily be determined. To prove this we use the differential geometry developed in the previous chapter. Liouville's Theorem allows us to study conformal geometry in space by considering the "space" of spheres and planes; this has its own metric structure which is related to the spacetime geometry of Special Relativity.

6.1 Conformal Mappings

There are mappings of Euclidean plane or space that preserve all the angles but not the distances; we have already learned about homotheties as an example. Since the angles determine the shape or the form of a rectilinearly bounded figure (independently of its size), the geometry of angles is also called *conformal geometry*. As in other domains of geometry, we will study in particular "isomorphisms", i.e. bijective angle-preserving mappings. We will see that these do not have to be linear, nor they must preserve straight lines. Nevertheless we want to study first all *linear* maps on Euclidean n-space $\mathbb{R}^n$ which preserve angles.

J.-H. Eschenburg, *Geometry – Intuition and Concepts*,
https://doi.org/10.1007/978-3-658-38640-5_6

Lemma 6.1 *A linear mapping $L : \mathbb{R}^n \to \mathbb{R}^n$ (with $n \geq 2$) preserves angles if and only if $\mu L \in O(n)$ for some $\mu > 0$.*

Proof Since the canonical basis vectors $e_1, \ldots, e_n$ include the angle 90° the same is true for their images $Le_1, \ldots, Le_n$. Moreover, from $e_i + e_j \perp e_i - e_j$ we have $0 = \langle Le_i + Le_j, Le_i - Le_j \rangle = |Le_i|^2 - |Le_j|^2$, hence all n vectors Le_i have the same length $|Le_i| = \lambda$. Therefore $(\mu Le_1, \ldots, \mu Le_n)$ with $\mu = \frac{1}{\lambda}$ is an orthonormal basis, i.e. (read as a matrix) an element of $O(n)$. □

What does angle preservation mean for *nonlinear* mappings? The maps are assumed to transform smooth curves into smooth curves, and the angle of intersection of two curves is supposed to be the same as the angle of intersection of their images. The first requirement suggests to consider differentiable mappings, more precisely C^1-*diffeomorphisms* (diffeomorphisms with continuous partial derivatives). A C^1-diffeomorphism $F : \mathbb{R}^n_o \to \mathbb{R}^n_1$ (two domains in $\mathbb{R}^n$) is called *angle preserving* or *conformal*, if two curves in $\mathbb{R}^n_o$ intersect at the same angle as their images under F: When $a, b : (-\epsilon, \epsilon) \to \mathbb{R}^n_o$ are regular curves (one-dimensional immersions) with $a(0) = b(0) = x$, then

$$\angle((Fa)'(0), (Fb)'(0)) = \angle(a'(0), b'(0)). \tag{6.1}$$

If one sets $a'(0) = v$ and $b'(0) = w$, then according to the chain rule $(Fa)'(0) = dF_x v$ and $(Fb)'(0) = dF_x w$. Thus the linear mapping dF_x must be angle preserving, and according to Lemma 6.1 we have for all $x \in \mathbb{R}^n_o$

$$dF_x \in \mathbb{R}^*_+ \cdot O(n), \tag{6.2}$$

or in other words, $dF_x/\lambda(x) \in O(n)$ for some $\lambda(x) > 0$, the *conformal factor*. The relation (6.2) can also be taken as a definition of conformality.

In the case $n = 2$, the group $\mathbb{R}^*_+ O(2)$ consists of two kinds of matrices: orientation preserving (positive determinant) $A = \left(\begin{smallmatrix} a & -b \\ b & a \end{smallmatrix}\right)$, and orientation reversing (negative determinant) $B = \left(\begin{smallmatrix} a & b \\ b & -a \end{smallmatrix}\right)$. When $\mathbb{R}^2$ is identified with $\mathbb{C}$, the first matrix A becomes multiplication by the complex number $a + ib$ because $(a + ib)(x + iy) = ax - by + i(ay + bx) = \left(\begin{smallmatrix} ax-by \\ ay+bx \end{smallmatrix}\right) = A\left(\begin{smallmatrix} x \\ y \end{smallmatrix}\right)$. A differentiable mapping on $\mathbb{R}^2 = \mathbb{C}$ whose Jacobian matrix at each point is the multiplication by a complex number is called *complex differentiable* or *holomorphic*.[1] By composing with a reflection, e.g., the complex conjugation $z \mapsto \bar{z}$, non-oriented conformal mappings pass into oriented ones; the non-oriented ones are thus holomorphic mappings followed by

[1] The usual definition is somewhat different: F is called *complex differentiable* on a domain $\mathbb{C}_o \subset \mathbb{C}$ if for every $z \in \mathbb{C}_o$ the limit $\lim_{h \to 0} \frac{F(z+h)-F(z)}{h} =: c$ exists, i.e. $O(h) := \frac{F(z+h)-F(z)}{h} - c \to 0$ for $h \to 0$. In other words, $F(z + h) - F(z) = ch + o(h)$ with $o(h) = hO(h)$ and $o(h)/|h| \to 0$, i.e., F is differentiable, and the derivative dF_z is multiplication by the complex number c.

complex conjugation *(antiholomorphic mappings)*. Thus, conformal geometry in dimension 2 is nothing but the theory of holomorphic functions in one complex variable. There is a tremendous variety of such functions; every power series represents a holomorphic function within its disk of convergence. We will see that for dimensions $n \geq 3$ the situation is completely different.

6.2 Inversions

Are there also nonlinear conformal mappings in higher dimensions? An important example is the *inversion*

$$F : \mathbb{R}^n_* \to \mathbb{R}^n_*, \quad F(x) = \frac{x}{|x|^2}. \tag{6.3}$$

It leaves the unit sphere pointwise fixed, $Fx = x$ for $|x| = 1$, and it is its own inverse mapping: $F^{-1} = F$ (such mappings are called *involutions*): $F(Fx) = Fx/|Fx|^2 = \frac{x}{|x|^2}/\frac{1}{|x|^2} = x$. So it is a kind of reflection along the unit sphere. To see its conformality we need to calculate its derivative: $dF_x v = \frac{d}{dt}F(x+tv)|_{t=0}$. Now $F(x+tv) = \frac{p(t)}{q(t)}$ with

$$p(t) = x + tv, \quad q(t) = \langle x + tv, x + tv \rangle\,,$$

thus $p(0) = x$, $p'(0) = v$ and $q(0) = |x|^2$, $q'(0) = 2\langle v, x\rangle$. Hence

$$\begin{aligned}
dF_x v &= ((p'q - pq')/q^2)(0) \\
&= \left(v|x|^2 - 2x\langle v, x\rangle\right)/|x|^4 \\
&= \frac{1}{|x|^2}\left(v - 2\left\langle v, \frac{x}{|x|}\right\rangle \frac{x}{|x|}\right) \\
&= S_x v/|x|^2,
\end{aligned}$$

where S_x denotes the reflection along the hyperplane $x^\perp$. So $dF_x \in \mathbb{R}^*_+ O(n)$ and F is conformal.

The inversion F has yet another property: It preserves the set of spheres and hyperplanes, and since a hyperplane can be considered as a sphere of radius ∞, we call this property *sphere preserving* or *spherical* for short. Indeed, spheres and hyperplanes are the solution sets of equations of the following type:

$$\alpha|x|^2 + \langle x, b\rangle + \gamma = 0 \tag{6.4}$$

where $\alpha, \gamma \in \mathbb{R}$ and $b \in \mathbb{R}^n$. Substituting $x = Fy$, we obtain

$$0 = \alpha\left|\frac{y}{|y|^2}\right|^2 + \left\langle\frac{y}{|y|^2}, b\right\rangle + \gamma = \frac{1}{|y|^2}(\alpha + \langle y, b\rangle + \gamma|y|^2)$$

hence $\alpha + \langle y, b\rangle + \gamma|y|^2 = 0$, and so y satisfies again an equation of type (6.4) where the roles of α and γ are interchanged.[2]

Because the inversion preserves spheres and fixes the unit sphere pointwise, it can be constructed geometrically, at least in dimension 2. From $|Fx| = 1/|x|$ we see $Fx \to 0$ for $|x| \to \infty$. Every straight line is therefore mapped into a circle or a straight line through 0, and the points of intersection with the unit circle remain fixed. The image of the straight line pq in the figure below is therefore the circle k through p, q and 0, and since the ray $0x$ is transformed into itself, F maps the intersection of $0x$ with pq to the intersection of $0x$ with k and vice versa. The triangle $(0, p, Fx)$ in the left figure is right-angled, and k is the Thales circle over the straight line segment $[0,x]$ in the right figure.

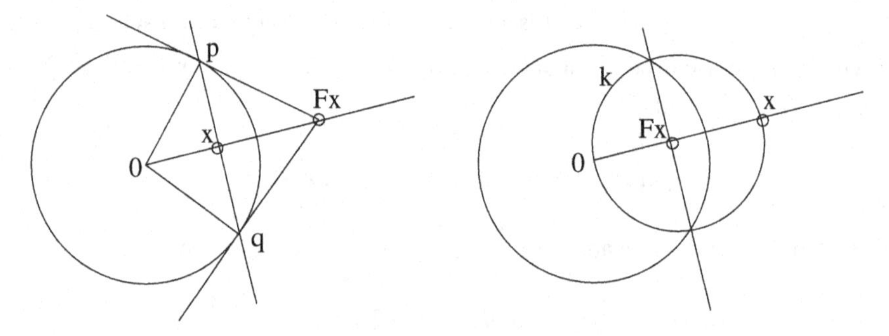

Analogously one defines an *inversion* F_S along an arbitrary sphere S: It fixes S and reflects all rays emanating from the center of S. If S has center 0 but arbitrary radius r, then F_S arises by conjugating F with the homothety S_r:

$$F_S(x) = rF(x/r) = \frac{r^2}{|x|^2}x. \tag{6.5}$$

The inversion along a sphere S with arbitrary center M is obtained by conjugation with the translation T_M, that is $F_S(x) = \frac{r^2}{|x-M|^2}(x - M) + M$. In each case the sphere S remains pointwise fixed, and M, x, F_Sx lie on a common line with

$$|x - M||F_Sx - M| = r^2. \tag{6.6}$$

[2] Actually, conformality follows from the sphere preserving property alone: Since the differential dF_x (as an approximation of F near x) is also sphere preserving and moreover linear, it transforms the unit sphere into a concentric sphere; such map preserves all norms up to one factor, thus it is a multiple of an orthogonal map.

6.3 Conformal and Spherical Mappings

Theorem 6.2 (Theorem of Liouville) [3] *Any conformal mapping $F : \mathbb{R}^n_o \to \mathbb{R}^n_1$ for $n \geq 3$ is also spherical. More precisely, F is a composition of inversions and hyperplane reflections.*

Proof We use the characterization of spheres and hyperplanes by the umbilic point property (Theorem 5.1). On spheres and hyperplanes we can build orthogonal hypersurface systems in any direction. The image of those under a conformal mapping is again an orthogonal hypersurface system; hence also on the image hypersurface, any tangential direction is a principal curvature direction.

In fact, let $S \subset \mathbb{R}^n$ be a sphere intersecting $\mathbb{R}^n_o$. We want to show that $\tilde{S} = F(S)$ is again an umbilic hypersurface. We move the center of S to the origin and choose spherical coordinates $\Phi : \mathbb{R}^n_2 \to \mathbb{R}^n_o$; by restricting Φ to a coordinate hyperplane, we obtain a parametrization φ of S. If $x = \varphi(u) \in S \cap \mathbb{R}^n_o$ and $v \in T_x S = x^\perp$ an arbitrarily given vector with $|v| = 1$, then by rotating the coordinate system we can obtain that $v = \frac{\partial \varphi}{\partial \theta}(u)$ where $u = (r, t_1, \ldots, t_{n-2}, \theta)$ (see Sect. 5.4). Since Φ is an orthogonal hypersurface system and F is angle-preserving, $\tilde{\Phi} = F \circ \Phi$ is again an orthogonal hypersurface system, and one of these hypersurfaces is $\tilde{S}$, parametrized by $\tilde{\varphi} = F \circ \varphi$. Thus the partial derivatives $\tilde{\varphi}_i = dF \cdot \varphi_i$ are principal curvature directions of $\tilde{\varphi}$. In particular $v = \frac{\partial \varphi}{\partial \theta}(u)$ and $dF_x v = \frac{\partial \tilde{\varphi}}{\partial \theta}(u)$ are principal curvature directions of φ and $\tilde{\varphi}$. Since v was arbitrary and dF_x maps the tangent hyperplanes isomorphically onto each other, *every* tangent vector of $\tilde{S}$ is a principal curvature direction.[4] Thus $\tilde{S}$ is an umbilic hypersurface and therefore an open part of a sphere or a hyperplane. Similarly, for a hyperplane H we can show that $F(H)$ is also an umbilic hypersurface. The mapping F is therefore spherical.

To prove the second assertion ("More precisely ..."), we again choose a point $p \in \mathbb{R}^n_o$ as well as sufficiently small spheres S_1 and S_2 with midpoints p and Fp. By the inversion F_1 along S_1, all (pieces of) spheres S through p are transformed into (pieces of) hyperplanes and vice versa. Similarly, the inversion F_2 along S_2 transforms spheres through Fp into hyperplanes and vice versa. Since F maps spheres through p onto spheres through Fp (here we count hyperplanes as spheres), $F' := F_2 \circ F \circ F_1$ maps hyperplanes into hyperplanes and therefore also straight lines into straight lines (being intersections of $n-1$ hyperplanes in general position); F' is therefore projective and at the same time conformal (angle preserving). But then parallels are also mapped onto parallels, because they lie in a common plane and have equal angles with a straight line intersecting them. Thus F' is linear and conformal, up to a translation, and hence it is the composition of an orthogonal map

[3] Joseph Liouville, 1809 (Saint Omer)–1882 (Paris).

[4] We have used here that the principal curvature directions on a hypersurface are independent of the parametrization. Indeed, we can use the isomorphism $d\varphi_u : \mathbb{R}^m \to T_u$ to view the Weingarten mapping A_u as a linear mapping from T_u instead of from $\mathbb{R}^m$; this is independent of the parametrization, cf. Sect. 5.2.

with a homothety. Translations and orthogonal maps are compositions of hyperplane reflections, and homotheties are concatenations of two inversions along concentric spheres, so $F = F_2 \circ F' \circ F_1$ is a concatenation of hyperplane reflections and inversions. □

6.4 The Stereographic Projection

Most of the conformal mappings just described cannot be defined on the entire $\mathbb{R}^n$: The inversion along a sphere S is not defined at the center of S, or in other terms, the center is mapped "to infinity". This was similar for projective mappings: Often, one hyperplane in affine space was mapped "to infinity" outside affine space. The solution to this problem in projective geometry was the extension of affine space to projective space by adding another hyperplane, the "hyperplane at infinity". In the same spirit we shall solve the problem in conformal geometry: We extend Euclidean space $\mathbb{R}^n$ to "conformal space", this time by adding just one new point, which we call "∞". But now the situation is simpler because we already know this extension: It is the n-dimensional sphere

$$\mathbb{S} := \mathbb{S}^n = \{(x, t) \in \mathbb{R}^n \times \mathbb{R} = \mathbb{R}^{n+1};\ |x|^2 + t^2 = 1\}, \tag{6.7}$$

and the embedding of $\mathbb{R}^n$ into $\mathbb{S}^n$ is done by *stereographic projection* $\Phi : \mathbb{R}^n \to \mathbb{S}^n$: Each point $x \in \mathbb{R}^n$ is connected by a straight line with the highest point of the sphere, the *north pole* $N = e_{n+1} = (0, 1)$. The image $\Phi(x) = (w, t)$ is the second intersection point of this straight line Nx with $\mathbb{S}^n$ (left figure).

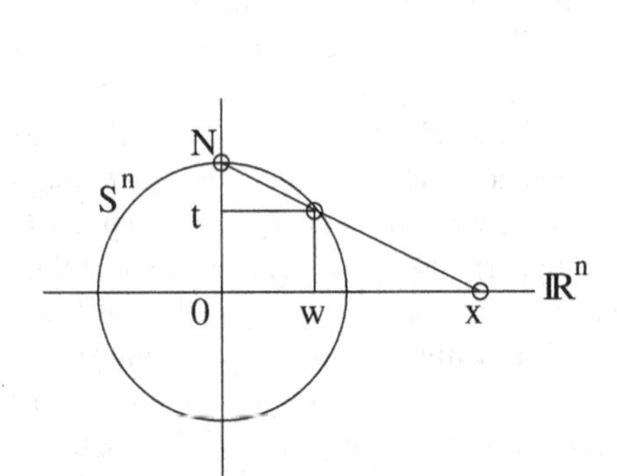

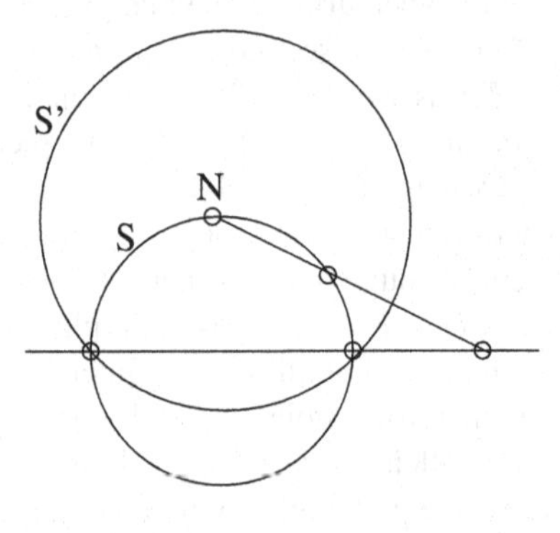

From the left figure we derive $x = \frac{w}{1-t}$ (similarity of the triangles $(x, 0, N)$ and $((w, t), (0, t), N)$); conversely, (w, t) is obtained from x by substituting $w \overset{*}{=} (1-t)x$ into the relation $|w|^2 + t^2 = 1$[5]:

$$\Phi : \mathbb{R}^n \to \mathbb{S}^n, \;\; \Phi(x) = \frac{1}{|x|^2+1}(2x, |x|^2-1), \;\; \Phi^{-1}(w,t) = \frac{w}{1-t}. \tag{6.8}$$

The right figure above shows that both Φ and Φ^{-1} are restrictions of the inversion $F_{S'}$ along the sphere S' through $\mathbb{S} \cap \mathbb{R}^n$ with center N. In fact, $F_{S'}$ maps the sphere $\mathbb{S}$ through the center N of S' onto the hyperplane through $S' \cap \mathbb{S}$ (which is $\mathbb{R}^n$), and the rays outgoing from N are kept invariant by $F_{S'}$. Thus both Φ and Φ^{-1} preserve angles and are spherical. This can also be seen directly from the following figures:

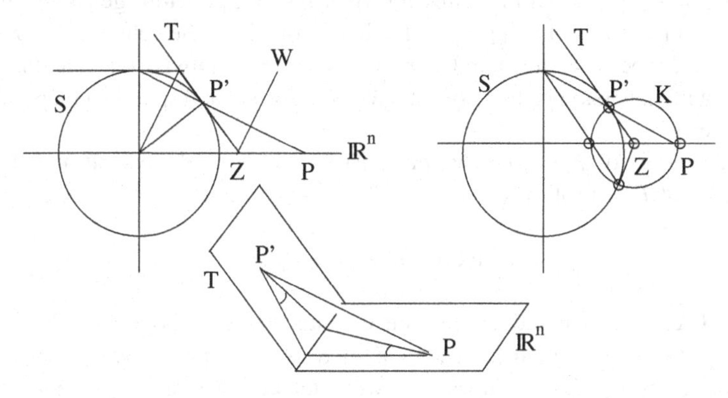

The projection line $P'P$ is perpendicular to the bisector W between the hyperplanes T (the tangent hyperplane of $\mathbb{S}$ at P') and $\mathbb{R}^n$; the reflection at W therefore transforms the angle between two tangents of $\mathbb{S}$ into the angle between their image lines under the central projection $\mathbb{R}^{n+1} \setminus \{N\} \to \mathbb{R}^n$ (projection center N) which extends the stereographic projection to the ambient space $\mathbb{R}^{n+1} \setminus \{N\}$. This shows the conformality of Φ (left and lower figures).

But the reflection along W also transforms the line segment $\overline{P'Z}$ onto the line segment $\overline{PZ}$; so these are of equal length (right figure). From this it follows that spheres are mapped onto spheres: A sphere k' in $\mathbb{S}$ through the point P' can be extended to a sphere $K \subset \mathbb{R}^{n+1}$, which intersects $\mathbb{S}$ perpendicularly along k'. The center of K is the point Z, the tip of the cone over k' tangent to $\mathbb{S}$. If necessary, we shift the image hyperplane $\mathbb{R}^n$ of Φ^{-1} in such a way that it becomes the horizontal hyperplane through Z, which changes the stereographic projection Φ^{-1} only by

[5] $1 = |w|^2 + t^2 \overset{*}{=} (1-t)^2|x|^2 + t^2 \iff (1-t)^2|x|^2 = 1 - t^2 = (1-t)(1+t) \overset{t<1}{\iff} (1-t)|x|^2 = 1 + t \iff |x|^2 - 1 = t(|x|^2+1) \iff t = (|x|^2-1)/(|x|^2+1)$ and $w = (1-t)x = 2x/(|x|^2+1)$.

composition with a homothety on $\mathbb{R}^n$. Since Z lies on the bisector W, the distance from Z to P' and P is the same, so both P and P' lie on the sphere K. Hence

$$\Phi^{-1}(\mathbb{S} \cap K) = \Phi^{-1}(k') = \mathbb{R}^n \cap K =: k. \tag{6.9}$$

Now let k be any (hyper-)sphere in $\mathbb{R}^n$ and F_k the inversion along k. How does F_k look like after transplanting it onto the sphere $\mathbb{S} = \mathbb{S}^n$ using Φ, i.e. passing to $\Phi \circ F_k \circ \Phi^{-1}$? The answer is given by the next figure. We extend k again to a sphere $K \subset \mathbb{R}^{n+1}$ which intersects both $\mathbb{S}$ and $\mathbb{R}^n$ (the image hyperplane of Φ^{-1} through Z) perpendicularly. By (6.9), $\mathbb{S} \cap K$ is the image sphere of $\mathbb{R}^n \cap K$ under the map $\Phi = F_{S'}|_{\mathbb{R}^n}$ (where $F_{S'}$ was defined after (6.8)). The inversion $F_{S'}$ thus transforms $\mathbb{R}^n$ to $\mathbb{S}$ and $\mathbb{R}^n \cap K$ to $\mathbb{S} \cap K$ since the sphere K (intersecting both $\mathbb{R}^n$ and $\mathbb{S}$ perpendicularly) remains invariant under $F_{S'}$. More generally, for any conformal and spherical mapping F, the inversion along the sphere $\tilde{K} = F(K)$ is $F \circ F_K \circ F^{-1}$ because this map leaves the sphere $\tilde{K}$ pointwise fixed, and it is easy to see that the inversion is the *only* nontrivial conformal and spherical mapping with this property.[6]

In particular for $F = F_{S'}$ we have $F(K) = K$, and since Φ and Φ^{-1} are restrictions of $F_{S'}$, it follows

$$\Phi \circ F_K \circ \Phi^{-1} = F_K|_{\mathbb{S}}. \tag{6.10}$$

Now F_K keeps invariant the sphere $\mathbb{S}$ since it intersects K orthogonally, and likewise any straight line g through the center Z of K is invariant. Hence $g \cap \mathbb{S}$ remains invariant. Thus F_K interchanges the two intersection points of g with $\mathbb{S}$. The mapping $\Phi \circ F_K \circ \Phi^{-1} = F_K|_{\mathbb{S}}$ thus causes exactly the interchange of these two points.[7]

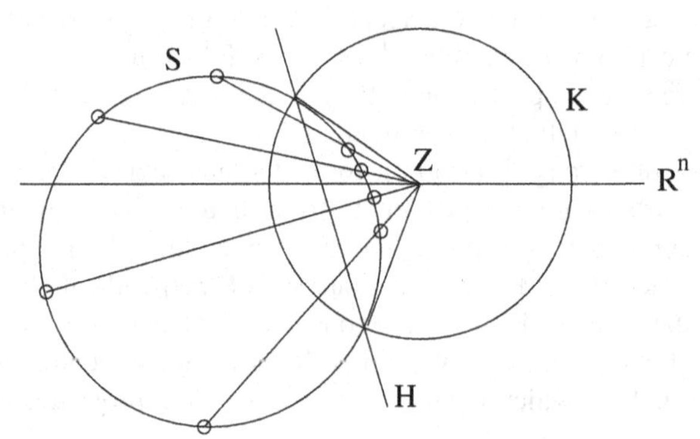

[6] If $F(K)$ is a hyperplane, $F \circ F_K \circ F^{-1}$ must be the reflection on this hyperplane for the same reason.

[7] This is at the same time another proof of the chord theorem for the circle; cf. Exercise 42, since the inversion F_K satisfies $|P - Z||F_K P - Z| = r^2$ where r is the radius of K.

6.5 The Space of Spheres

Our conformal mappings on the unit sphere $\mathbb{S} = \mathbb{S}^n$ keep the set of spheres in $\mathbb{S}$ invariant.[8] Each such sphere can be described as the intersection of $\mathbb{S}$ with a hyperplane $H \subset \mathbb{R}^{n+1}$ or equivalently with a sphere $K \subset \mathbb{R}^{n+1}$ which intersects $\mathbb{S}$ perpendicularly. The center Z of K is the tip of the tangent cone over $\mathbb{S} \cap H$, in other words, the *pole* of the hyperplane H, see Sect. 3.8 and Exercise 26. The pole determines uniquely both the hyperplane H and the sphere $\mathbb{S} \cap H$; the set of spheres in $\mathbb{S}$ can therefore be regarded as the set of possible poles, i.e. points of $\mathbb{R}^{n+1}$ outside of $\mathbb{S}$. But if H passes through the origin, $\mathbb{S} \cap H$ is a great sphere and the pole is moved to infinity. Thus we must extend $\mathbb{R}^{n+1}$ to $\mathbb{RP}^{n+1}$, and we better describe $\mathbb{S}$ as a quadric in $\mathbb{RP}^{n+1}$, viz.

$$\mathbb{S} = \{[x] \in \mathbb{RP}^{n+1};\ \langle x, x\rangle_- = 0\}, \tag{6.11}$$

where

$$\langle x, y\rangle_- = \sum_{i=1}^{n+1} x_i y_i - x_{n+2} y_{n+2} \text{ for all } x, y \in \mathbb{R}^{n+2}$$

is the *Lorentzian scalar product*[9] on $\mathbb{R}^{n+2}$. The space $\mathcal{K}$ of all spheres in $\mathbb{S}$ (more precisely: their poles) is the outer space of $\mathbb{S}$ in $\mathbb{RP}^{n+1}$. This is the set of *spacelike*[10] homogeneous vectors:

$$\mathcal{K} = \{[x] \in \mathbb{RP}^{n+1};\ \langle x, x\rangle_- > 0\}. \tag{6.12}$$

Let $O(n+1, 1)$ be the group of all linear maps on $\mathbb{R}^{n+2}$ which keep this scalar product invariant; it is called *Lorentz group*[11]:

$$A \in O(n+1, 1) \iff \langle Ax, Ay\rangle_- = \langle x, y\rangle_- \text{ for all } x, y \in \mathbb{R}^{n+2}.$$

[8] This holds for *invertible* conformal maps f even for $n = 2$: Possibly after composition with complex conjugation we may assume that f is holomorphic on its domain and meromorphic when defined on all of $\hat{\mathbb{C}} = \mathbb{S}^2$. Moreover, when f is injective it can have only one pole, a simple one. Thus it is *fractional-linear*, also called Möbius transformation, that is $f(z) = \frac{az+b}{cz+d}$, and hence it is a composition of isometries and inversions.

[9] Hendrik Antoon Lorentz, 1853 (Arnhem)–1928 (Haarlem).

[10] A vector $v \in \mathbb{R}^{n+2}$ is called *spacelike* if $\langle v, v\rangle_- > 0$. The notion comes from the interpretation of the Lorentzian scalar product as spacetime-metric in Einstein's theory of Special Relativity, see next footnote.

[11] The Lorentzian scalar product and the Lorentz group play an important role in Einstein's Special Relativity. Einstein implemented the observation that the speed of light c is the same in all uniformly moving systems. Even if we try to chase after a beam of light, it will move away from us always at the same relative speed c. A light wave emanating from the origin at time 0 forms at time t the sphere with the equation $x_1^2 + x_2^2 + x_3^2 = x_4^2$ where $x_4 := ct$. In $\mathbb{R}^4$ = space × time these spheres together form the *light cone* $C = \{x \in \mathbb{R}^4;\ \langle x, x\rangle_- = 0\}$. It has the same equation in any uniformly moving coordinate system. The corresponding coordinate transformations essentially form the Lorentz group.

The corresponding projective mappings on $\mathbb{RP}^{n+1}$ leave both $\mathbb{S}$ and $\mathcal{K}$ invariant, and they are spherical on $\mathbb{S}$ (hence conformal, see Footnote 2): A linear map $A \in O(n+1,1)$ maps a sphere $H \cap \mathbb{S}$ (where H is a hyperplane meeting $\mathbb{S}$) onto the sphere $A(H) \cap \mathbb{S}$.

In particular, the *Lorentzian reflection*[12] along such hyperplane H restricts on $\mathbb{S}$ to the *inversion* along the sphere $\mathbb{S} \cap H$ because it fixes $\mathbb{S} \cap H$ (see also the figure after Eq. (6.10)). Since these Lorentzian hyperplane reflections generate the group $O(n+1,1)/\{\pm I\}$, as the Euclidean hyperplane reflections generate $O(n)$ (cf. Theorem 4.3 in Sect. 4.4), and since on the other hand the inversions generate the group $\mathrm{Conf}(\mathbb{S}^n)$ of the conformal mappings on $\mathbb{S}^n$ (the *Möbius group*)[13] as we have seen in Theorem 6.2 (Liouville), the two groups are identical:

$$\mathrm{Conf}(\mathbb{S}^n) = PO(n+1,1) = O(n+1,1)/\pm\,. \tag{6.13}$$

From this point of view, conformal geometry (actually the geometry of spheres) is a subdomain of projective geometry on $\mathbb{RP}^n$, more precisely the *polar geometry* because there is a *polarity* which is preserved under the admissible transformations: the Lorentzian scalar product $\langle\ ,\ \rangle_-$.

6.6 Möbius and Lie Geometry of Spheres

The Möbius geometry of spheres in the sphere $\mathbb{S} = \mathbb{S}^n$ as described above can be refined by replacing homogeneous vectors $[x] \in \mathcal{K} \subset \mathbb{RP}^{n+1}$ with their generating vectors $x \in \mathbb{R}^{n+2}$ for which $\langle x, x\rangle_- = 1$. In each $[x]$ there are two such vectors $x, -x$; thus each sphere $K \subset \mathbb{S}^n$ is represented twice. Geometrically, the doubling corresponds to the two possible orientations or sides of the sphere. Unlike spheres in $\mathbb{R}^n$, spheres in $\mathbb{S}^n$ have no "inside" and "outside", but both sides are completely equal (like northern and southern hemisphere at the equator). The homogeneous vector $[x]$ defining a sphere $K = \mathbb{S} \cap H$ as its pole is Lorentz-perpendicular to the hyperplane H defining K. The two vectors $\pm x$ mark the two sides of both the hyperplane H and the sphere $K = \mathbb{S} \cap H$. The *oriented spheres* thus form a quadric in $\mathbb{R}^{n+2}$,

$$L = \{x \in \mathbb{R}^{n+2};\ \langle x, x\rangle_- = 1\},$$

[12] If $H \subset \mathbb{R}^{n+2}$ is a hyperplane meeting $\mathbb{S}$, its Lorentzian normal space $H^\perp = \{v \in \mathbb{R}^{n+1};\ \langle h, v\rangle_- = 0\ \forall h \in H\}$ lies outside $\mathbb{S}$ (more precisely, $[H^\perp] \cap \mathbb{S} = \emptyset$), it is spacelike. Thus $\mathbb{R}^{n+2} = H \oplus H^\perp$. The Lorentzian reflection R_H along H keeps this decomposition invariant with $R_H = I$ on H and $R_H = -I$ on $H^\perp$.

[13] August Ferdinand Möbius, 1790 (Pforta)–1868 (Leipzig).
In fact, a spherical mapping F on $\mathbb{R}^n$ is a composition of hyperplane reflections and inversions, cf. Theorem 6.2. Thus $\tilde F = \Phi \circ F \circ \Phi^{-1}$ on $\mathbb{S}^n$ is a composition of inversions since this holds whenever F is an inversion or a hyperplane reflection.

a Lorentz analogue of the unit sphere in Euclidean space $\mathbb{R}^{n+2}$. It is a *Lorentzian submanifold* of $\mathbb{R}^{n+2}$, that is the inner product $\langle\ ,\ \rangle_-$ on $\mathbb{R}^{n+2}$ restricts on each tangent space $T_xL = x^\perp$ to a Lorentzian inner product again. The Lorentz group $O(n+1,1)$ acts on L transitively and isometrically. Hence the geometry of the Lorentzian manifold L is just the Möbius geometry of the space of *oriented* spheres.

We can go one step further and add to the oriented spheres of arbitrary radius the "spheres of radius zero", the *points*. This step was taken by Sophus Lie,[14] and this geometry was named Lie sphere geometry to his honor. Algebraically, one has to pass to the projective closure of the quadric L: As explained in Sect. 3.6, one has to add another coordinate x_{n+3} and "homogenize" the equation $\langle x,x\rangle_- = 1$ to $\langle x,x\rangle_- = x_{n+3}^2$ or $\langle \hat{x},\hat{x}\rangle_= = 0$ for $\hat{x} = (x_1,\dots,x_{n+3})$, where

$$\langle \hat{x},\hat{x}\rangle_= := x_1^2 + \dots + x_{n+1}^2 - x_{n+2}^2 - x_{n+3}^2.$$

This scalar product is neither Euclidean nor Lorentzian, because it has *two* minus signs ("index 2"). The projective closure of L is thus

$$\hat{L} = \{[\hat{x}] \in \mathbb{RP}^{n+2};\ \langle \hat{x},\hat{x}\rangle_= = 0\},$$

and the invariance group $O(n+1,2)$ of the scalar product $\langle\ ,\ \rangle_=$ acts transitively on it (actually $PO(n+1,2)$). The "spheres of radius zero" (points) are represented by the points at infinity of $\hat{L}$, the intersection of $\hat{L}$ with the hyperplane at infinity $\mathbb{RP}^{n+1}$ (last coordinate $x_{n+3} = 0$), and $\hat{L} \cap \mathbb{RP}^{n+1}$ is the sphere $\mathbb{S}^n$, the set of points in $\mathbb{S}^n$. The Lie sphere geometry is therefore the polar geometry with polarity $\langle\ ,\ \rangle_=$.

[14] Sophus Lie, 1842 (Nordfjordeid, Norway)–1899 (Christiania, Oslo), 1886–1898 in Leipzig.

Angular Distance: Spherical and Hyperbolic Geometry

7

Abstract

The geometry of the sphere is familiar to us, from everyday life as well as from geography. It is a part of the metric geometry of space, yet it represents something of its own in it. The distance of two points on the unit sphere is its angle, measured from the center; thus the angle takes on a whole new meaning: spherical distance. There is a second geometry which is similarly defined, but has exactly opposite properties in many respects: Here the surrounding Euclidean space is replaced by $\mathbb{R}^{n+1}$ with the Lorentzian scalar product, the spacetime of Special Relativity. The "unit sphere" in this space is a model of the non-Euclidean geometry of Lobachevski and Bolyai, which had caused a great surprise in the early nineteenth century because it contradicted the common belief that Euclidean geometry was the only conceivable geometry.

7.1 Hyperbolic Space

In Sect. 6.5 we have studied the action of the *Lorentz group* $PO(n, 1)$ on the projective space $\mathbb{RP}^n$. It has three orbits:

1. The sphere $\mathbb{S}^{n-1} = C/\mathbb{R}^*$, where $C = \{x \in \mathbb{R}^{n+1}_*;\ \langle x, x\rangle_- = 0\}$ is the *light cone* which in Special Relativity describes the tangent vectors of light rays,
2. The unit ball $B^n = \{[x] \in \mathbb{RP}^n;\ \langle x, x\rangle_-. < 0\} = \{[v, 1];\ v \in \mathbb{R}^n,\ |v| < 1\}$,

J.-H. Eschenburg, *Geometry – Intuition and Concepts*,
https://doi.org/10.1007/978-3-658-38640-5_7

3. The outer space $\mathbb{RP}^n \setminus \bar{B}^n$ where $\bar{B}^n \subset \mathbb{R}^n \subset \mathbb{RP}^n$ is the closed unit ball, more precisely

$$\begin{aligned}\bar{B}^n &= \{[v, 1] \in \mathbb{RP}^n;\ v \in \mathbb{R}^n,\ |v| \le 1\}\\ &= \{[x] \in \mathbb{RP}^n;\ \langle x, x\rangle_- \le 0\},\\ \mathbb{RP}^n \setminus \bar{B}^n &= \{[x] \in \mathbb{RP}^n;\ \langle x, x\rangle_- > 0\}.\end{aligned}$$

On the sphere $\mathbb{S}^{n-1}$ the Lorentz group acts by conformal spherical transformations, as we have seen.[1] We had identified the outer space as the space of spheres in $\mathbb{S} = \mathbb{S}^{n-1}$ by assigning to each sphere in $\mathbb{S}$ (intersection with an affine hyperplane) its *pole* (tip of the corresponding tangent cone). The spherical mappings on $\mathbb{S}$ transform the space of spheres in $\mathbb{S}$ accordingly. It remains to study the interior space B^n. This represents something new: a model of *hyperbolic geometry*, and the Lorentz group $PO(n, 1)$ is the associated isometry group. To introduce this geometry, we consider a shell of the two-sheeted hyperboloid[2] sitting in the light cone C:

$$H = \{x \in \mathbb{R}^{n+1};\ \langle x, x\rangle_- = -1,\ x_{n+1} > 0\}. \tag{7.2}$$

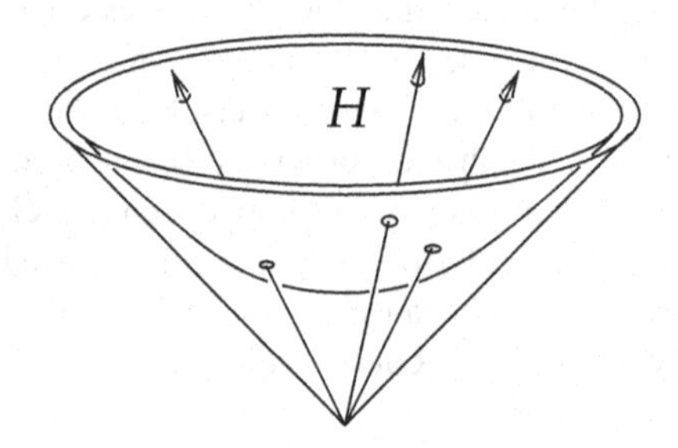

We can use H to represent the points in $B^n \subset \mathbb{RP}^n$, which are the homogeneous vectors $[x]$ with $\langle x, x\rangle_- < 0$: In each class $[x]$ with $\langle x, x\rangle_- < 0$ we find exactly one x with $\langle x, x\rangle_- = -1$ and $x_{n+1} > 0$; the one-dimensional subspace $[x]$ intersects the shell H of the hyperboloid exactly once. The Lorentzian scalar product of the surrounding space $\mathbb{R}^{n+1}$ defines angles and distances on the hypersurface H,

[1] Specifically on $\mathbb{S}^2 = \mathbb{C} \cup \{\infty\} = \hat{\mathbb{C}} = \mathbb{CP}^1$, the group of oriented conformal circle-preserving transformations is the group of fractional-linear functions $f(z) = \frac{az+b}{cz+d}$ with $ad - bc \neq 0$. This in turn is the projective group in complex dimension 1, $PGL(2, \mathbb{C})$. Two-dimensional oriented conformal geometry and one-dimensional complex projective geometry are thus identical,

$$PO^+(3, 1) = PGL(2, \mathbb{C}) \tag{7.1}$$

(where $^+$ stands for orientation-preserving). Therefore, we can also describe the Lorentz transformations of Special Relativity by complex 2×2-matrices. Eq. (7.1) is one of the coincidences between low-dimensional Lie groups, of which there are several more (e.g. $SU(2) = \mathbb{S}^3 \subset \mathbb{H}$).

[2] In *projective* space $\mathbb{RP}^n$, the two shells are identified.

just like the Euclidean scalar product. Although the Lorentzian scalar product is indefinite, its restriction to any tangent space $T_x H$ of H is again positive definite. This is clear for $x = e_{n+1}$ because the tangent hyperplane is "horizontal" and has no x_{n+1}-component where the scalar product is negative, $T_{e_{n+1}} H = \mathbb{R}^n \subset \mathbb{R}^{n+1}$. And it is also true on $T_x H$ for any $x \in H$ because the slope of $T_x H$ is always smaller than that of the light cone, i.e. for a Lorentzian normal vector $v \perp T_x H$ the "timelike" component v_{n+1} predominates. Or even easier, any $x \in H$ can be transformed to e_{n+1} by a Lorentz transformation[3] $A \in O^+(n, 1)$ preserving H (because H is defined through the scalar product preserved by A), so the tangent hyperplane of H at x is mapped onto that at e_{n+1}, and the Lorentzian scalar product on $T_x H$ is transformed into the one on $T_{e_{n+1}} H = \mathbb{R}^n$ which is positive definite. The hyperboloid shell with its geometry induced by the Lorentzian scalar product will be called *hyperbolic space* or for $n = 2$ *hyperbolic plane*.

7.2 Distance on the Sphere and in Hyperbolic Space

In order to understand formally and also intuitively how the scalar product of the surrounding space defines a geometry on the hypersurface, let us first consider a simpler example: the sphere $\mathbb{S} = \mathbb{S}^n = \{x \in \mathbb{R}^{n+1};\ \langle x, x\rangle = 1\}$ in $\mathbb{R}^{n+1}$ with the ordinary Euclidean scalar product. We have already talked about angles: The angle between two curves on $\mathbb{S}$ intersecting each other in $x \in \mathbb{S}$ is defined by the angle of their tangent vectors in $T_x\mathbb{S}$ which in turn is defined by the scalar product of the surrounding space $\mathbb{R}^{n+1}$, restricted to $T_x\mathbb{S}$. But what about distances? After all, we all live on a sphere, on the surface of our planet Earth, hence we are familiar with the problem from everyday life. The distance between two points $x, y \in \mathbb{S}$ could still be defined as $|x - y|$ but this notion of distance would not be very practical: It measures the length of the line segment between x and y that passes through the Earth. An airplane from x = London to y = Tokyo will not be able to take that route. The distance from x to y that has meaning for a traveler is the length of a connecting curve on the Earth's *surface* that is as short as possible:

$$|x, y| := \inf\{L(c);\ c : x \overset{\mathbb{S}}{\rightsquigarrow} y\}. \tag{7.3}$$

Here $c : x \overset{\mathbb{S}}{\rightsquigarrow} y$ means a C^1-curve $c : [a, b] \to \mathbb{R}^{n+1}$ with $c(a) = x$, $c(b) = y$ and $c(t) \in \mathbb{S}$ for all $t \in [a, b]$, and $L(c)$ is the *arc length* of c

$$L(c) = \int_a^b |c'(t)|\, dt. \tag{7.4}$$

[3] The subset $\{x \in \mathbb{R}^{n+1};\ \langle x, x\rangle < 0\}$ of *timelike* vectors has two connected components: $x_{n+1} > 0$ and $x_{n+1} < 0$. By $O^+(n, 1)$ we denote the subgroup of $O(n, 1)$ preserving each connected component. Apparently this is isomorphic to $PO(n, 1) = O(n, 1)/\{\pm I\}$.

We all know from everyday experience (or a compactness argument) that this infimum is a minimum: There exists a shortest curve from x to y on $\mathbb{S}$. This is a *great circle* arc which is defined as follows. Every plane E intersects $\mathbb{S}$ in a circle (unless $E \cap \mathbb{S} = \emptyset$), and if the plane passes through O (the center of $\mathbb{S}$), that is, if $E \subset \mathbb{R}^3$ is a two-dimensional linear subspace, it meets $\mathbb{S}$ where it is "thickest", in a circle of maximal radius (which is the radius 1 of $\mathbb{S}$). This is called a *great circle*. The geographic latitude of London is 51.5° north, that of Tokyo is less than 36°, but still an airplane from London to Tokyo on the shortest route starts almost northbound, following a great circle arc. We can see easily that the great circle arc between x and y is shortest when we use spherical coordinates, but rotated such that x and y are lying on a common meridian (then that great circle arc is a subset of this meridian).

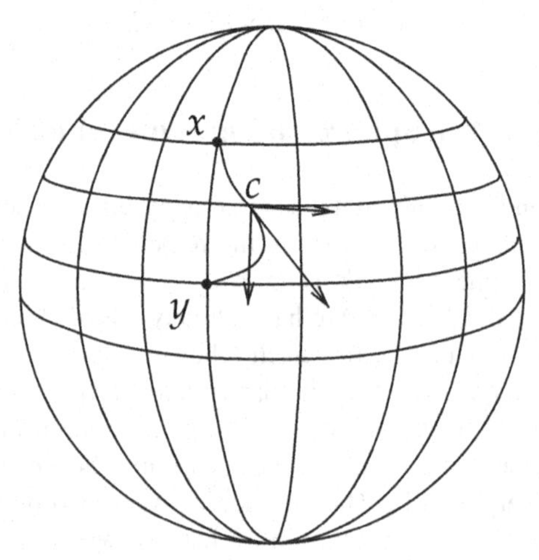

In these coordinates, y is situated precisely south of x. Each curve $c : x \rightsquigarrow y$ deviating from the meridian is longer since the velocity vector (derivative) c' also has an east-west component, while the north-south component alone gives (integrated) the latitude difference and thus the length of the meridian; because the east-west component is added, the length of c is greater than the latitude difference between x and y.[4] We can see this analytically using the formula for the spherical coordinates φ, θ on $\mathbb{S}^2$ given in (5.14); then $c(t)$ is expressed by two functions $\theta(t)$, $\varphi(t)$:

$$c(t) = \begin{pmatrix} \sin\theta(t)\cos\varphi(t) \\ \sin\theta(t)\sin\varphi(t) \\ \cos\theta(t) \end{pmatrix},$$

[4] The corresponding argument does not hold when instead x and y lie on a common *circle of latitude*, because on the northern hemisphere these circles get shorter towards the north. Thus swerving to the north still costs an extra north-south component, but at the same time the length of the east-west component is diminished.

and the derivative of this is

$$c'(t) = \theta'(t) \begin{pmatrix} \cos\theta(t)\cos\varphi(t) \\ \cos\theta(t)\sin\varphi(t) \\ -\sin\theta(t) \end{pmatrix} + \varphi'(t) \begin{pmatrix} -\sin\theta(t)\sin\varphi(t) \\ \sin\theta(t)\cos\varphi(t) \\ 0 \end{pmatrix}.$$

The two vectors on the right-hand side[5] are orthogonal with norm 1 and $\sin\theta$, therefore

$$|c'| = \sqrt{(\theta')^2 + (\varphi')^2 \sin^2\theta} \geq \theta',$$

hence the length satisfies $L(c) = \int_a^b |c'(t)|\,dt \geq \int_a^b \theta'(t)dt = \theta(b) - \theta(a)$. Of course, $\mathbb{S}^2 \subset \mathbb{R}^3$ can be replaced by $\mathbb{S}^n \subset \mathbb{R}^{n+1}$.

The situation is similar if we replace the unit sphere $\mathbb{S}^n$ in Euclidean space by the hyperbolic space which is the "unit sphere" in Lorentzian space or the hyperboloid shell $H \subset \mathbb{R}^{n+1}$, see (7.2). As in Euclidean case we conclude that the shortest curves (the "hyperbolic lines") are the intersections $H \cap E$ of H with two-dimensional linear subspaces $E \subset \mathbb{R}^{n+1}$ that intersect H. But now the arc length of curves $c : [a, b] \to H$ is measured using the *Lorentzian* scalar product: We still have (7.4) with $|c'| = \sqrt{\langle c', c'\rangle_-}$. The argument is the same as for the sphere; the analogue of the spherical coordinates for $n = 2$ is the mapping

$$(\varphi, \theta) \mapsto \begin{pmatrix} \sinh\theta\cos\varphi \\ \sinh\theta\sin\varphi \\ \cosh\theta \end{pmatrix},$$

where the angular functions for θ have been replaced by the hyperbolic functions sinh, cosh; this mapping is a parametrization of H since $\cosh^2 - \sinh^2 = 1$. The arc length of a hyperbolic straight line is also called the *hyperbolic angle* by analogy with the spherical case. As in the spherical case, by applying a Lorentz transformation (an element of the group $O(n, 1)$) we can make x and y lie on a common meridian.

---

[5] These are the two partial derivatives of the map $(\varphi, \theta) \mapsto \begin{pmatrix} \sin\theta\cos\varphi \\ \sin\theta\sin\varphi \\ \cos\theta \end{pmatrix}$.

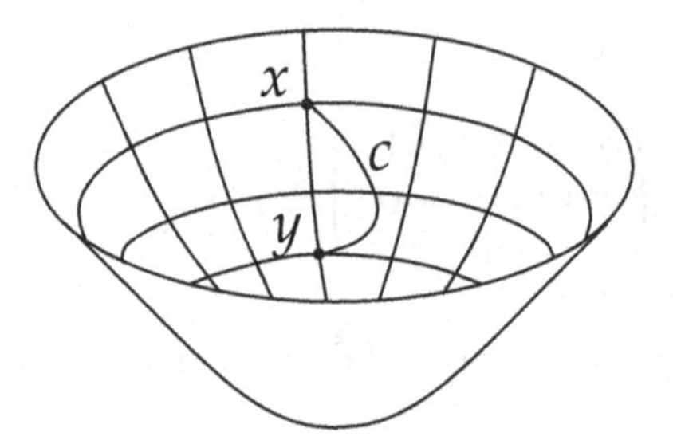

7.3 Models of Hyperbolic Geometry

If we project H by straight lines from the origin 0 onto the horizontal tangent hyperplane at the point e_{n+1}, we obtain an open disk D, and the hyperbolic straight lines $H \cap E$ are mapped onto straight line segments $D \cap E$ within D.

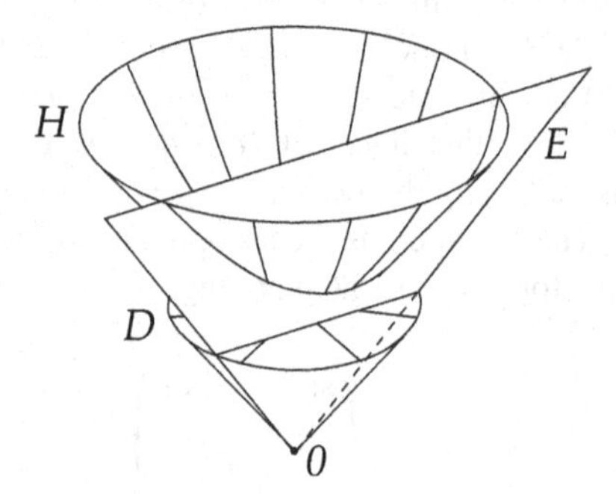

This is the model of hyperbolic geometry named after Felix Klein: As a set it is the open unit ball D, and the hyperbolic straight lines are just the straight line segments within D. It is a model of *non-Euclidean geometry* developed by *Lobachevski*, *Bolyai*,[6] and Gauss which satisfies all axioms of geometry established by Euclid with the exception of the axiom of *parallels*: For a straight line g and a point P outside g in hyperbolic plane, there is no longer only one, but a great many straight lines h which do not meet g and are "parallel" to g in that sense.

[6] Nikolai Ivanovich Lobachevski, 1792 (Nizhny Novgorod)–1856 (Kazan), János Bolyai, 1802 (Clausenburg/Cluj Napoca)–1860 (Neumarkt/Târgu Mureş, now Romania).

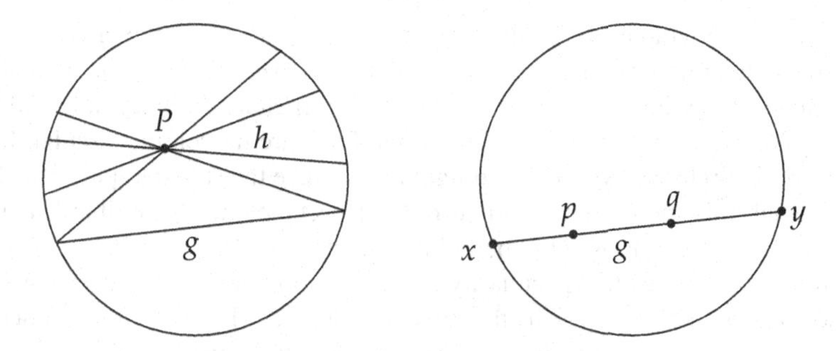

Also the hyperbolic distance (the hyperbolic angle) $d_h(p,q)$ between two points $p, q \in H$ can be read from Klein's model (right figure):

$$d_h(p,q) = \frac{1}{2}|\log|(p,q;x,y)||. \tag{7.5}$$

Here $x, y \in S = \partial D$ are the intersection points of the connecting line $g = pq$ with the boundary of D, and $(\, , \; ; \, , \,)$ denotes the cross-ratio introduced in Sect. 3.10, see Exercise 45.

The boundary of the disk D^n is the sphere $\mathbb{S}^{n-1}$ on which the Lorentz group $PO(n,1)$ acts by conformal transformations. Considered as a subgroup of $PO(n+1,1)$ (the conformal group on $\mathbb{S}^n$), it consists of those conformal maps on $\mathbb{S}^n$ which preserve the upper hemisphere $\mathbb{S}^n_+$ and thus also the equator $\mathbb{S}^{n-1}$. This consideration leads us to a second model of hyperbolic geometry which is named after *Henri Poincaré*.[7] Instead of the straight line segments g in D^n we consider the small half circles on $\mathbb{S}^n_+$, which lie vertically above g and which are orthogonally projected onto g (left figure):

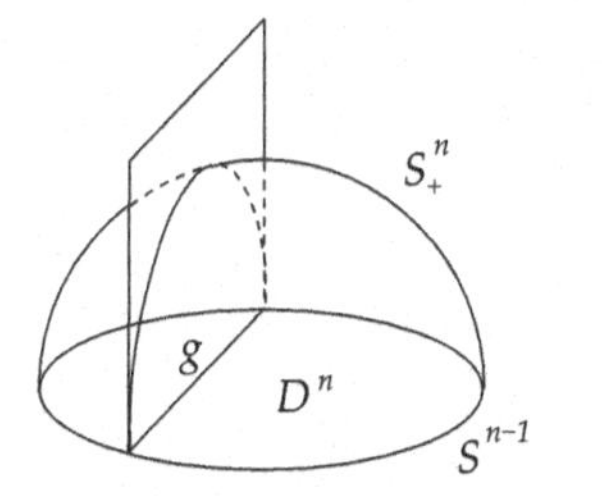

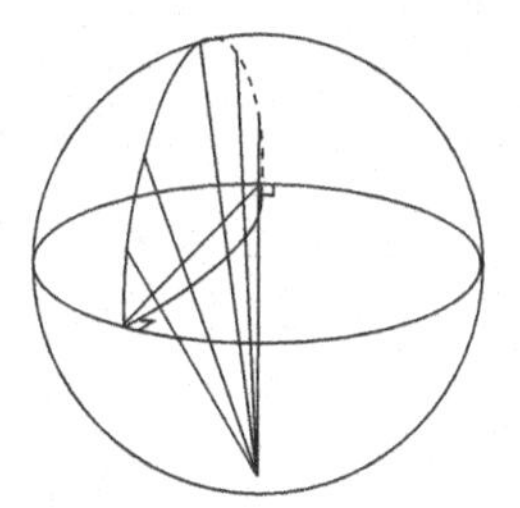

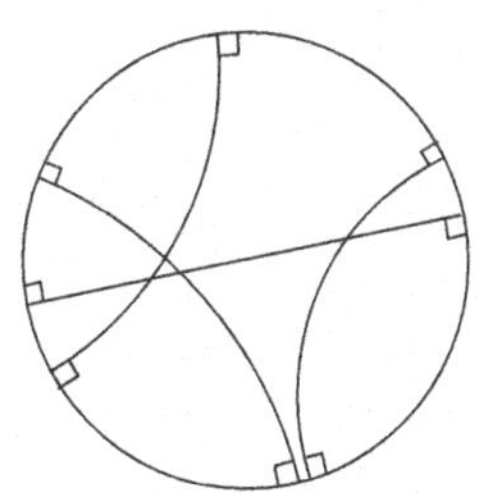

All these circular arcs intersect the equator $\mathbb{S}^{n-1}$ perpendicularly. Since the subgroup $PO(n,1) \subset PO(n+1,1)$ keeps the equator and the upper hemisphere $\mathbb{S}^n_+$ invariant and maps circles to circles, the circles perpendicular to the equator are mapped to just such circles. Since these transformations also preserve angles, the Euclidean angles which we see are the true angles also in hyperbolic geometry.[8]

[7] Jules Henri Poincaré, 1854 (Nancy)–1912 (Paris).

[8] For this argument, we need a point where the angles of Euclidean and hyperbolic geometry already coincide; this is the point e_{n+1} because on the (horizontal) tangential hyperplane $T_{e_{n+1}}H =$

If we apply stereographic projection from the south pole of $\mathbb{S}^n$, the circular arcs in $\mathbb{S}^n_+$ perpendicular to the equator are mapped into the *orthocircles* in D, circular arcs which intersect the boundary sphere $\mathbb{S}^{n-1}$ of D perpendicularly (middle and right figures). The right figure reproduces the usual Poincaré model: the circular disk D with the orthocircles as hyperbolic straight lines. Since the stereographic projection is conformal, this model is conformal too: At every point, the visible Euclidean angles are the same as those of hyperbolic geometry.

In dimension 3, one more peculiarity occurs: In Footnote 8, Chap. 6 we have seen that the conformal group on $\mathbb{S}^2$ is the group of fractional-linear complex functions, which are the projective mappings of the complex projective line $\mathbb{CP}^1 = \hat{\mathbb{C}}$ (see Exercise 15 and Footnote 1)). Thus we have $PO(3,1)^\circ = PGL(2,\mathbb{C})^\circ$, where $^\circ$ stands for the connected component of the group. The complex projective geometry in complex dimension 1 coincides with the two-dimensional conformal geometry. This has implications for hyperbolic geometry in dimension 3 since this is the conformal geometry of the $\mathbb{S}^2$. We have therefore seen:

Theorem 7.1 *The isometry group of the hyperbolic space H^3, the group of Möbius transformations $PGL(2,\mathbb{C})$, and the Lorentz group $PO(3,1)$ of Special Relativity have isomorphic connected components.*

$\mathbb{R}^n$ the Lorentzian scalar product coincides with the Euclidean one. The group $PO(n,1)$ operates transitively on $\mathbb{S}^n_+$ and on H. It preserves the (Euclidean) angles on $\mathbb{S}^n_+$ because it operates conformally, and it preserves the angles of hyperbolic geometry on H because it operates isometrically on H (since it preserves the Lorentzian scalar product and thus the hyperbolic distance).

Exercises

8

Abstract

The exercises are a carefully selected supplement to the material. They offer the opportunity to pursue special situations or questions for which there is not enough space in the systematic presentation. They vary in difficulty, so hints are often given to facilitate access.

Exercise 1 *"The moon has risen"*
Consider the following three sketches.

(a) In which direction is the sun?
(b) What can you tell about the daytime in each case? (Day, night, morning, evening, before or after sunrise or sunset?) Please give reasons!

8.1 Affine Geometry (Chap. 2)

Exercise 2 *Group actions*
Let X be a set and $(G, \cdot)$ a group. An *action* of G on X is a mapping $w : G \times X \to X$, $(g, x) \mapsto w_g x$ with the properties $w_e = \mathrm{id}_X$ for the neutral element $e \in G$ and $w_{gh} = w_g w_h$ for all $g, h \in G$ (cf. Sect. 2.2). Show:

J.-H. Eschenburg, *Geometry – Intuition and Concepts*,
https://doi.org/10.1007/978-3-658-38640-5_8

(a) The relation $\sim$ on X, defined by $x \sim y \iff_{\text{Def}} \exists_{g \in G} : y = w_g x$ is an *equivalence relation*: $x \sim x$; $x \sim y \Rightarrow y \sim x$; $x \sim y,\ y \sim z \Rightarrow x \sim z$. The equivalence classes $[x] = \{w_g x;\ g \in G\} =: Gx$ are called *orbits* of the action w.

(b) For each $x \in X$ the subset $G_x := \{g \in G;\ w_g x = x\}$ is a subgroup of G (called *stabilizer* or *isotropy group* of x).

(c) By G/G_x we denote the set of cosets: A *coset* of G_x is a subset of G of the form $gG_x = \{gh;\ h \in G_x\}$ for some $g \in G$, and $G/G_x := \{gG_x;\ g \in G\}$. Show that the mapping $w^x : G/G_x \to [x] = Gx$ with $gG_x \mapsto w_g x$ is well-defined and bijective. (Thus, the orbit type does not depend on the action w, but only on the isotropy group).

Exercise 3 *Cube group*

Let X be the set of vertices of a cube in coaxial position (i.e. edges parallel to the three coordinate axes) and G the set of all rotations of the cube that render X into a coaxial position again. Determine the number of elements (*cardinality*) of G using the following considerations: Any vertex of the cube can be rotated into any other, and there are three rotations each of which leaves a particular vertex fixed (why three?). What is the connection to Exercise 2?

Exercise 4 *Pappus' theorem and commutativity*

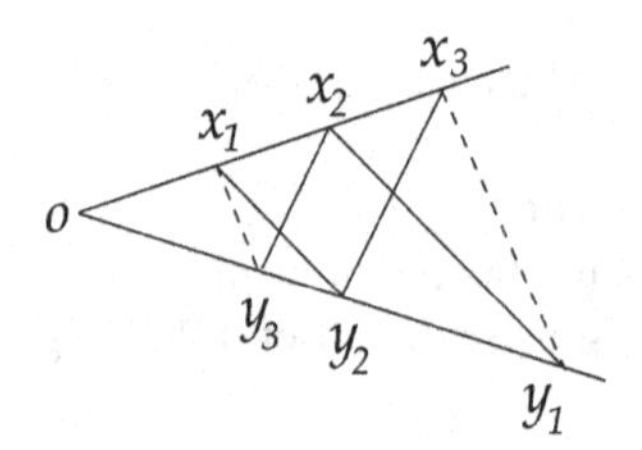

Let X be a vector space over a field or skew field $\mathbb{K}$. Given two straight lines through o and on each of it three points x_1, x_2, x_3 and y_1, y_2, y_3. Assume that the pairs of lines x_1y_2, x_2y_1 and x_2y_3, x_3y_2 are parallel. Show that then always also the third (dashed in the figure) pair of lines x_1y_3, x_3y_1 is parallel *(Pappus' theorem)*, provided that $\mathbb{K}$ is commutative, and vice versa: If this property is always satisfied, then $\mathbb{K}$ is commutative.

Hint: Use the geometric characterization of the homothety S_λ, cf. Exercise 5.

Exercise 5 *Homotheties and Desargues' theorem*

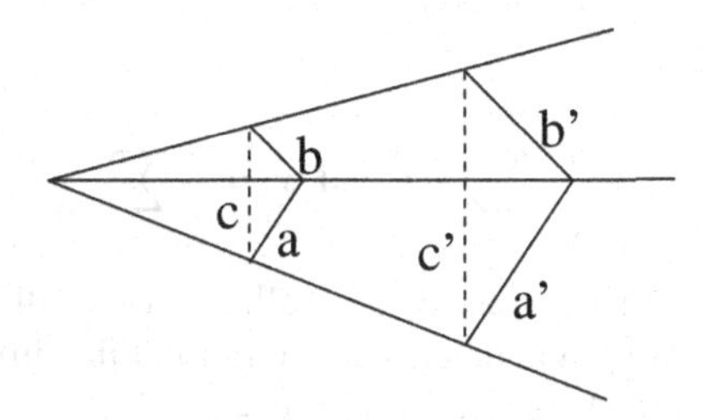

The homothety $S_\lambda : x \mapsto \lambda x$ can be geometrically characterized as follows: If $x, \lambda x, y \in X$ are given, then λy is the intersection of the line oy with the line parallel to xy through the point λx. Show: The fact that S_λ preserves directions, $[\forall_{x,y \in X}\ S_\lambda(xy) \,||\, xy]$, is geometrically translated into the *theorem of Desargues* (Chap. 3.5):

> *Given two triangles with vertices on three straight lines through o such that two of the three edge pairs are parallel: $a \,||\, a'$ and $b \,||\, b'$, then the third pair of edges (dashed in the figure) is also parallel: $c \,||\, c'$.*

Exercise 6 *Affine group* Aff(X)
An affine map F on a vector space X has the form $F(x) = Ax + a$ for some linear map A on X and some $a \in X$; more precisely, the correspondence between affine maps F and pairs (A, a) is bijective. Show that the invertible affine maps on a vector space X form a group, called the *affine group* Aff(X). Compute the pair corresponding to the composition of two invertible affine mappings F, G on X which correspond to the pairs (A, a) and (B, b). Moreover, compute the pair of the inverse map F^{-1}.

Exercise 7 *Medians (gravity lines) of a simplex*
Let X be an n-dimensional real (i.e. $\mathbb{K} = \mathbb{R}$) affine space and $a_0, ..., a_n$ affinely independent points. The *n-simplex*[1] Σ spanned by $a_0, \dots, a_n$ is the convex hull of these points:

$$\Sigma = \langle a_0, \dots, a_n \rangle = \left\{ \sum\nolimits_j \lambda_j a_j;\ \sum\nolimits_j \lambda_j = 1,\ 0 \le \lambda_j \le 1 \right\}.$$

[1] An n-simplex for $n = 0$ is a point, for $n = 1$ a *line segment*, for $n = 2$ a *triangle*, for $n = 3$ a *tetrahedron*.

The *sides* of Σ are the $(n-1)$-simplices $\Sigma_i = \langle a_0, \dots \hat{a}_i \dots, a_n \rangle$ for $i = 0, \dots, n$ where $\hat{a}_i$ denotes the *absence*, the *omission* of a_i. By s and s_i we denote the *centers of gravity* of Σ and Σ_i,

$$s = \frac{1}{n+1} \sum_j a_j \text{ and } s_i = \frac{1}{n} \sum_{j \neq i} a_j.$$

Show that the three points a_i, s and s_i are collinear (we call the common straight line *gravity line* or *median*), and determine the ratio of the difference vectors $(a_i - s)/(s_i - s)$. Please draw a sketch (figure) for $n = 2$.

Exercise 8 *Medians of a triangle*

(a) Show that each median is decomposing a triangle into two parts of equal area (area of a triangle = $\frac{1}{2}$ · base · height).
(b) Is this also true for any other straight line passing through the center of gravity? Does it split the triangle into two equal-area parts?
Proof or counterexample!

Exercise 9 *The Euler line*[2]

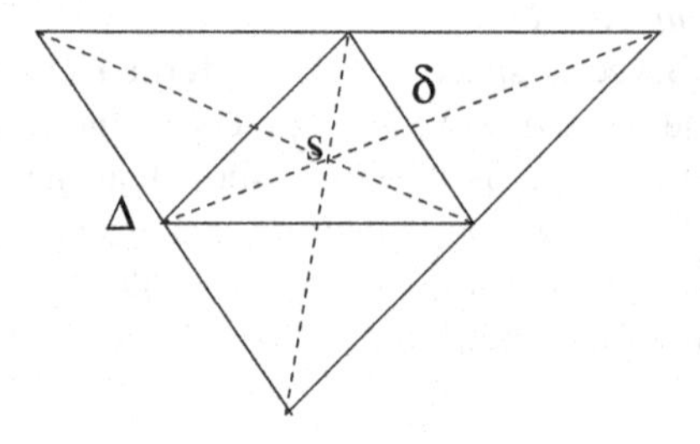

Consider a triangle δ and the circumscribed triangle Δ with parallel edges as in the figure. The points related to δ will be denoted by lower case letters.

(a) Show that the two triangles have a common center of gravity (*centroid*) s.
(b) Show that Δ arises from δ by applying a homothety with center $o = s$ and scaling factor $\lambda = -2$.
(c) Show that the altitudes[3] of δ are the perpendicular bisectors[4] of Δ.

[2] Leonhard Euler, 1707 (Basel)–1783 (St. Petersburg).

[3] An *altitude* of a triangle is a straight line through a vertex of the triangle which is perpendicular to its opposite edge. The exercise also shows that the three altitudes meet in a common point, the *orthocenter* of the triangle.

[4] A *perpendicular bisector* for a triangle is a line perpendicular to an edge through the midpoint of that edge. The three perpendicular bisectors meet in the *circumcenter*, the center of the circle passing through all three vertices (*circumcircle*) of Δ. Proof: Each perpendicular bisector has equal distance from two vertex point. The circumcenter has equal distance from all three vertices.

(d) Conclude that the center of gravity (*centroid*) s, the center of the circumcircle (*circumcenter*) m and the intersection point of the altitudes (*orthocenter*) h of the triangle δ lie on a common straight line. What is the ratio $(m - s)/(h - s)$?

Exercise 10 *Parallel projection of octahedron and icosahedron*

(a) (See parallel projection figure at the beginning of Sect. 4.5.) Draw the *octahedron* with the vertices $\pm e_1, \pm e_2, \pm e_3$ in space $\mathbb{R}^3$ (where $e_1 = (1, 0, 0)$, $e_2 = (0, 1, 0)$ etc.). Use for projection the affine mapping $F : \mathbb{R}^3 \to \mathbb{R}^2$ with $F(0) = 0$, $F(e_1) = (-5, -4)$, $F(e_2) = (6, -3)$, $F(e_3) = (0, 6)$. Into this octahedron draw an *icosahedron* whose 12 vertices lie on the 12 edges of the octahedron, subdividing those at the ratio of the *golden section.* [5]
In this way, each side of the octahedron is inscribed with a smaller equilateral triangle, and the subdivision point on each edge of the octahedron determines these smaller triangles in both octahedral triangles adjacent to the edge.

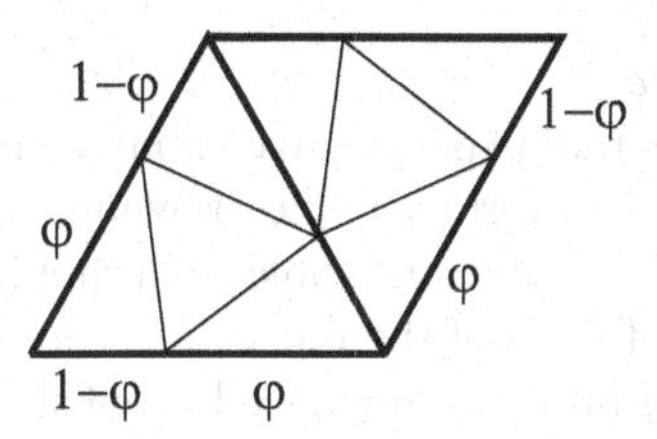

Let the subdivision point on the edge e_1e_3 be closer to e_3 than to e_1. Note that ratios of parallel line segments are preserved under the parallel projection (why?). Imagine the octahedron composed of rods and the inscribed icosahedron as a solid body. Draw only visible edges, and clarify the crossings with octahedral edges (which of them is above?).

(b) This icosahedron has two sorts of edges: One lies on an octahedron side, the other connects points on adjacent octahedron sides. Show that nevertheless all edges have the same length.

(c) Show that this construction is given also by the vertices of the symbol of the Matheon at Berlin (www.matheon.de), where three congruent “golden”

[5] The *golden section*, see Exercise 28, divides a line segment of length $a + b$ into two parts a and b with $a > b$ such that $a/b = (a + b)/a$. The ratio $\Phi := a/b$ thus satisfies $\Phi = 1 + 1/\Phi$ or $\Phi^2 = \Phi + 1$, thus $\Phi = \frac{1}{2}(\sqrt{5} + 1)$. Its inverse b/a is called φ. This ratio φ is approximated by the quotients of successive *Fibonacci numbers* 5/8, 8/13, 13/21 etc. The *Fibonacci numbers* (named after Leonardo of Pisa, called Fibonacci, c. 1170–1240) form the sequence of numbers beginning with 0 and 1, where each following number is the sum of its two predecessors: 0, 1, 1, 2, 3, 5, 8, 13, 21, ...

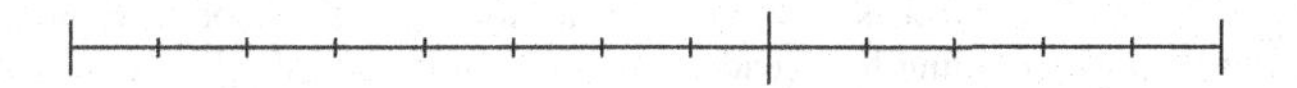

rectangles (i.e. whose side lengths are in the golden ratio) are stuck together using a "golden" slit in the middle of each rectangle, see figures.

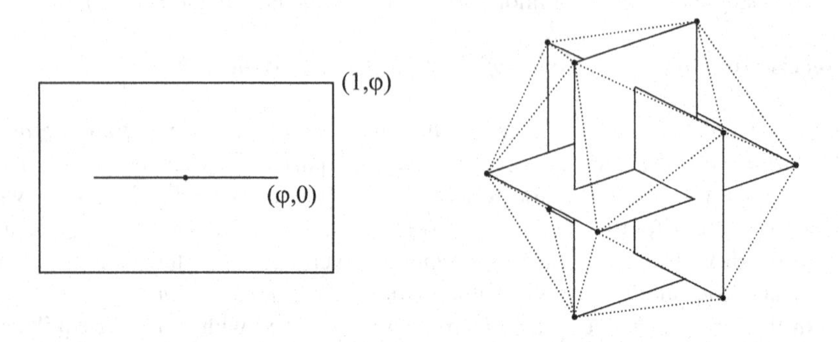

8.2 Projective Geometry (Chap. 3)

Exercise 11 *Perspective*
Draw a house with gable roof in perspective obliquely from above (bird's eye view). Vertical lines should also be vertical in the drawing, and the right edge of the gable front should be in front. In reality, the gable front should be square up to the base of the roof, and the slope of the roof should be 45°. Your drawing should also include the horizons of all planes involved (floor, walls, roof slopes). Give brief explanations and justifications for all construction steps.

Exercise 12 *Photographic image*
Determine in three-dimensional space (coordinates x, y, z) the central projection with projection center (0, 1, 1), which maps the horizontal xy-plane onto the vertical xz-plane (draw a figure!), i.e. calculate the assignment rule $(x, y) \mapsto (x', z')$. Determine the horizon in the xz-plane and show that the images of parallel straight lines of the xy-plane have an intersection on the horizon (except for the lines parallel to the horizon).

Exercise 13 *Dürer's "Saint Jerome"*
Please consider closely the copperplate engraving "St. Jerome in his Study" ("Der Heilige Hieronymus im Gehäus") by Albrecht Dürer (1514).[6] It contains a wealth of geometric ideas. With the additional knowledge that some angles are right angles in reality (table, window, ...) you can construct the exact 3-dimensional model of the room from the perspective representation! With the help of the figure below explain

[6] E.g. see image at https://en.wikipedia.org/wiki/Saint_Jerome_in_His_Study (Dürer).
Jerome, 347 (Stridon, Dalmatia)–414 (Bethlehem), translated the New Testament from Greek into Latin (Vulgate): "Gloria in excelsis Deo, et in terra pax hominibus bonae voluntatis." He was reported animal loving; according to legend, he pulled a thorn out of a lion's paw, so he is often depicted with the lion.

some of its elements: The table is "in reality" square and the angle between the edge and the diagonal of the chair (f and g) is "in truth" the angle β between f' and g' in the figure.

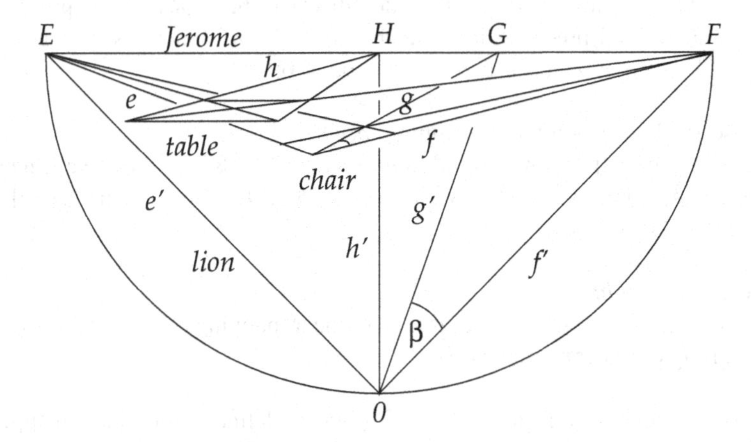

Exercise 14 *Projective mappings*

(a) A Discussion:

A: A projective mapping of the real plane should be defined by the images of three points. It comes, after all, from a linear map of $\mathbb{R}^3$, and this is known to be defined by the images of three basis vectors. Three points specify a plane; a table with three legs is known not to tilt.

B: But we have seen in perspective images that we can map a square by a projective mapping onto any quadrilateral. Thus we should be able to specify four points, not just three!

C: Is this true about the *arbitrary* quadrilateral? Doesn't the image of the (convex) square under a projective mapping have to be convex again? Convexity is defined using straight lines, after all, and straight lines map back to straight lines. But not all quadrilaterals are convex!

You: Who do you agree with and who do you disagree with? With which arguments? (See part (b).)

(b) Consider the example of the projective mapping $F = [A]$ where A maps the basis vectors $b_1 = (1, 0, 1)$, $b_2 = (0, 1, 1)$ and $b_3 = (0, 0, 1)$ onto b_1, b_2 and $-b_3$. What is the image of the square with vertices $[0, 0, 1]$, $[1, 0, 1]$, $[0, 1, 1]$, $[1, 1, 1]$ in the affine plane $\mathbb{A}^2 = \{[x, y, 1];\ x, y \in \mathbb{R}\} \subset \mathbb{P}^2$? Evaluate the preceding discussion using this result. Who is right? Where are there errors in reasoning?

(c) Try to formulate and prove a general result of the following type: "*A projective mapping of n-dimensional projective space $\mathbb{P} = \mathbb{P}^n$ over any field $\mathbb{K}$ is determined by the images of any k points in $\mathbb{P}$* (specify k as a function of n) *with the following properties ...*" Start with the case $n = 2$.

Exercise 15 *Projective line*
The projective line $\mathbb{P}^1 = P_{\mathbb{K}^2}$ over any field $\mathbb{K}$ can be identified with $\hat{\mathbb{K}} := \mathbb{K} \cup \{\infty\}$ using the mapping $\phi : \hat{\mathbb{K}} \to \mathbb{P}^1$ with $\phi(x) = [x, 1]$ for all $x \in \mathbb{K}$ and $\phi(\infty) = [1, 0]$. Show that with this identification the projective mappings from $\mathbb{P}^1$ are transformed precisely into the *fractional-linear* functions $f : \hat{\mathbb{K}} \to \hat{\mathbb{K}}$, $f(x) = \frac{ax+b}{cx+d}$ for $a, b, c, d \in \mathbb{K}$ with $ad - bc \neq 0$.

Exercise 16 *Which semilinear mappings are trivial on* $\mathbb{P}$?
Let V be a vector space over an arbitrary (skew) field $\mathbb{K}$. Determine all invertible semilinear mappings $S : V \to V$ which act on P_V as the identity, that is $[Sv] = [v]$ for all $v \in V_* = V \setminus \{0\}$.

Exercise 17 *Semilinear group*
Let $\Gamma L(\mathbb{K}^n)$ be the group of invertible semilinear mappings on $\mathbb{K}^n$ and $GL(\mathbb{K}^n)$ be the subgroup of linear mappings. Show:

(a) Any field automorphism $\sigma : \mathbb{K} \to \mathbb{K}$, $\lambda \mapsto \lambda^\sigma$ defines a semilinear mapping on $\mathbb{K}^n$, viz. $x = (x_1, \dots, x_n) \mapsto x^\sigma := (x_1^\sigma, \dots, x_n^\sigma)$. Thus the group Aut($\mathbb{K}$) of all field automorphisms becomes a subgroup of the semilinear group $\Gamma L(\mathbb{K}^n)$ which intersects the linear group $GL(\mathbb{K}^n)$ only at the identity id.
(b) Show that any semilinear mapping in $\Gamma L(\mathbb{K}^n)$ can be written uniquely as a composition of a linear mapping with a field automorphism.
(c) Compute this composition for a product $A\alpha B\beta$ with $A, B \in GL(\mathbb{K}^n)$ and $\alpha, \beta \in \mathrm{Aut}(\mathbb{K})$ and show that $\Gamma L(\mathbb{K}^n)$ is a semidirect product of $G = GL(\mathbb{K}^n)$ and $H = \mathrm{Aut}(\mathbb{K})$.
Reminder: A semidirect product[7] *of two groups G and H is the Cartesian product $G \times H$ with the group multiplication*

$$(g, h)(g', h') = (gw_h(g'), hh'),$$

where w is an action of H on G by automorphisms, i.e. a group homomorphism $w : H \to \mathrm{Aut}(G)$. *As an example or model you already know the affine group which is a semidirect product of the translation group $G = \mathbb{R}^n$ and the linear group $H = GL(n, \mathbb{R})$, see* Exercise 6.

Exercise 18 *Collineations preserve subspaces!*
Let V be a vector space over some field $\mathbb{K}$ and $P = P_V = \{[v];\ v \in V \setminus \{0\}\}$ the corresponding projective space. Show geometrically that a collineation $F : P_V \to P_V$ maps every k-dimensional projective subspace of P_V back to a k-dimensional subspace.
Hints: Induction over k. A k-dimensional subspace is defined by a $(k-1)$-dimensional subspace and a straight line.

[7] https://en.wikipedia.org/wiki/Semidirect_product.

Exercise 19 *Circle, parabola, hyperbola are projectively equivalent!*
Homogenize the equations of the circle $x^2 + y^2 = 1$, the parabola $y = x^2$ and the hyperbola $x^2 - y^2 = 1$. Give explicitly invertible linear mappings on $\mathbb{R}^3$ defining projective mappings of the $\mathbb{RP}^2$ which transform the projective closures of the circle, parabola and hyperbola into each other.

Exercise 20 *Projective mappings*
(See Exercise 14) Show: Every projective mapping $F = [A] : \mathbb{P}^n \to \mathbb{P}^n$ is defined by $n+2$ points in $\mathbb{P}^n$, e.g. the images of the points $[e_1], \ldots, [e_{n+1}]$ and $[e]$ with $e := e_1 + \ldots + e_{n+1} = \sum_i e_i$ (where e_i for $i = 1, \ldots, n+1$ are the canonical basis vectors of the $\mathbb{K}^{n+1}$ with $e_1 = (1, 0, \ldots, 0)$, $e_2 = (0, 1, 0, \ldots, 0)$,..., $e_{n+1} = (0, \ldots, 0, 1)$). Thereby the image points $[a_i] = F[e_i] = [Ae_i]$ and $[a] = F[e] = [Ae]$ in $\mathbb{P}^n$ can be given arbitrarily under the condition that every subset of $n + 1$ elements of $\{a_1, \ldots, a_{n+1}, a\}$ is linearly independent.
Hints: Choose $n + 2$ such vectors $a_1, \ldots, a_{n+1}, a$, then $a = \sum_i \mu_i a_i$ with known $\mu_i \in \mathbb{K}$, and all $\mu_i \neq 0$ (why?). Now we can set $Ae_i = \lambda_i a_i$ and $Ae = \lambda a$ and compute the still unknown scalars λ_i and λ (except for a common factor $\neq 0$) from the relations between e and e_i and between a and a_i.

Exercise 21 *Points at infinity on the hyperbola*
Explain why in the case of the hyperbola $\{(x, y) \in \mathbb{R}^2;\ x^2 - y^2 = 1\}$ the points at infinity of the asymptotes $x = y$ and $x = -y$ (draw a figure!) must be regarded as the two points at infinity of the hyperbola.[8] Explain the facts in as many ways as possible: geometrically and algebraically, in the plane and with the help of the cone $\{x^2 - y^2 = z^2\}$ in space.
Note: Please write as you would like it explained to yourself! Don't be afraid to recall "familiar" things: What are points at infinity? How is the projective plane explained? What is the role of the cone with the equation $x^2 - y^2 - z^2 = 0$? What does it look like? Draw figures.

Exercise 22 *Projective type of a quadric*
Determine the normal form (the type) of the projective quadric $Q \subset \mathbb{RP}^3$, which is the solution set of the following equation:

$$x^2 + 2y^2 + z^2 + 4w^2 + 4xy + 6xw + 8yw - 2zw = 0.$$

[8] In fact, the projective closures of the asymptotes are the *tangents* (see Exercise 23) of the projective closures of the hyperbola in these two points.

Exercise 23 *Straight lines on the one-sheeted hyperboloid*
Consider the one-sheeted hyperboloid $Q = \{[s, t, u, v] \in \mathbb{P}^3;\ st = uv\}$. Show that the straight lines

$$g_\lambda = \{[s, t, u, v];\ s/u = v/t = \lambda\},$$
$$h_\mu = \{[s, t, u, v];\ s/v = u/t = \mu\}$$

for any fixed $\lambda, \mu \in \mathbb{K} \cup \{\infty\}$ are lying completely on the surface Q and on its tangent planes at all points of Q through which the lines pass.
Recall the terms tangent, tangent plane, tangent space: The tangent space of the quadric $Q = \{[x] \in \mathbb{P}^n;\ \beta(x, x) = 0\}$ *at the point* $[x] \in Q$ *is the hyperplane* $T_{[x]}Q = \{[v] \in \mathbb{P}^n;\ \beta(x, v) = 0\}$. *Here* β *denotes any symmetric bilinear form on* $\mathbb{K}^{n+1}$.

Exercise 24 *Pappus' theorem, projective version*
Formulate (draw a figure!) and prove the projective version of Pappus' theorem (cf. Exercise 4) in a projective plane over a (commutative) field. What is the projective analogue of the parallels that occurred in the affine version (which is assumed to be known)?

Exercise 25 *Dual theorem for Pappus*
Determine the dual theorem for the projective theorem of Pappus (cf. the following figures) and demonstrate its statement by means of a figure of your own.

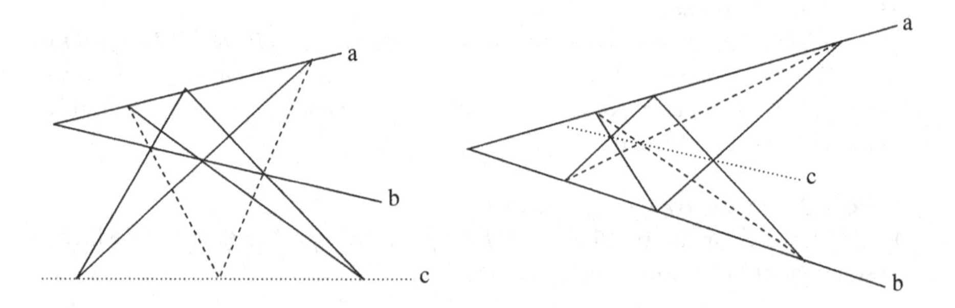

Additional question: Pappus' theorem is Pascal's theorem for a *degenerate* conic section, a pair of straight lines (see the second figure on Pascal's theorem, Sect. 3.8); it is thus a limiting case of Pascal's theorem. Can we also interpret the dual theorem as a limiting case of Brianchon's theorem?

Exercise 26 *Polarity*
Consider the polarity on $\mathbb{RP}^2$ which is defined by the bilinear form $\beta(v, w) = v_1w_1 + v_2w_2 - v_3w_3 = \langle v, w\rangle_-$ on $\mathbb{R}^3$: To any point $P = [x] \in \mathbb{RP}^2$ (the *pole*) we assign the straight line $g = \{[v];\ \beta(x, v) = 0\} \subset \mathbb{RP}^2$ (the *polar*), and vice versa. Show that this polarity, restricted to the affine plane, is given by the construction

below (see figures). Use that the polarity interchanges the operations "intersecting" and "connecting".

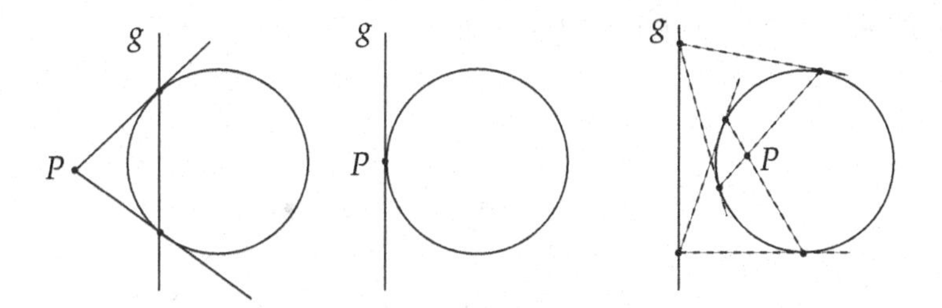

Exercise 27 *Cross-ratio*
Show: The projective mappings on $\mathbb{P}^1$ are exactly the bijective mappings $F : \mathbb{P}^1 \to \mathbb{P}^1$ which leave the cross-ratio invariant, that is

$(Fx, Fy; Fz, Fw) = (x, y; z, w)$ for all $x, y, z, w \in \mathbb{P}^1 = \mathbb{K} \cup \{\infty\}$.

8.3 Euclidean Geometry (Chap. 4)

Exercise 28 *Pentagon and golden section*

(a) Show for the regular pentagon (see figure) that the diagonal $a + b$ and the side length a are in the golden ratio: $\frac{a+b}{a} = \frac{a}{b}$. Use the similarity (?!) of the hatched triangles in the following figure.

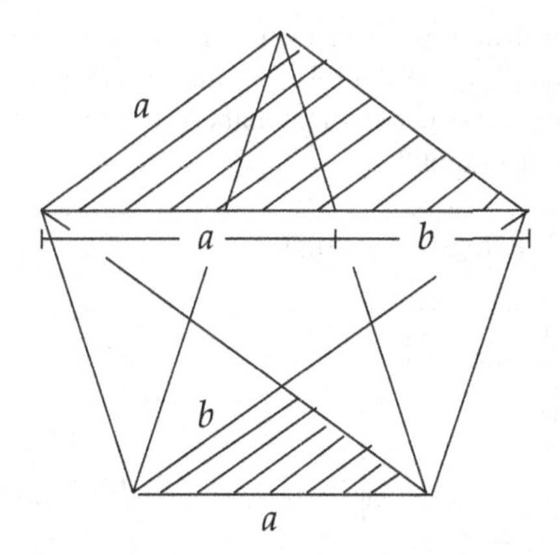

(b) The golden section $\frac{a}{b} > 1$ is sometimes denoted by the Greek letter Φ (Phi), after the ancient Greek sculptor Phidias (ca. 500–430 BC), who used it many times in his art. Because of $\frac{a}{b} = \frac{a+b}{a} = 1 + \frac{b}{a}$ we have $\Phi = 1 + 1/\Phi$ or $\Phi^2 = \Phi + 1$, from which we obtain $\Phi = \frac{1}{2}(\sqrt{5}+1)$. The reciprocal $\frac{b}{a} = 1/\Phi < 1$ will be denoted by φ, and from $\Phi = 1 + 1/\Phi$ it follows that $\varphi = \Phi - 1$. Multiplying this equation by φ we obtain $\varphi^2 = 1 - \varphi$. Clear in every detail?

Exercise 29 *Dürer's pentagon construction*

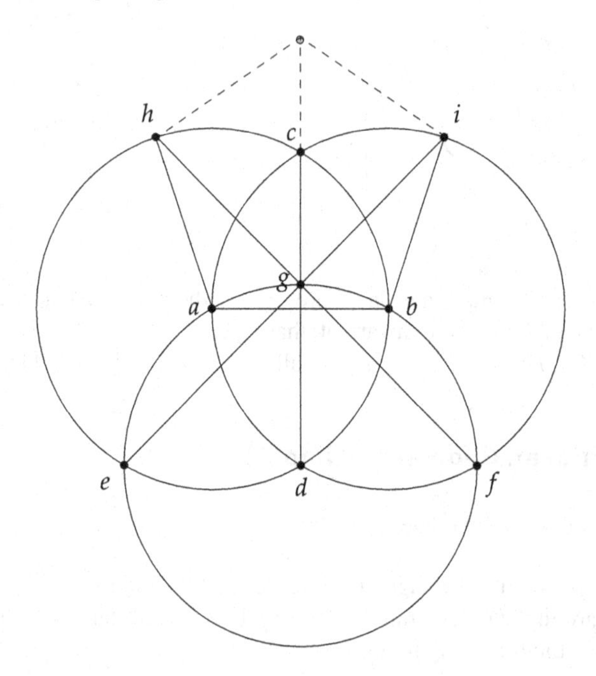

Is the pentagon construction given in the figure from Albrecht Dürer's "Underweysung der Messung" [8] exact? The points a and b are given, and the points c, d, e, f, g, h, i are constructed in alphabetical order. Are the points h, a, b, i vertices of a regular pentagon?

Exercise 30 *Usual pentagon construction*

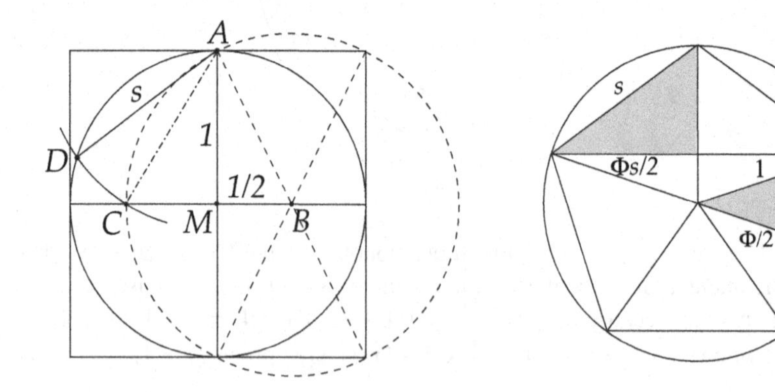

Show that the line segment $\overline{AD}$ in the left figure is a side of the regular pentagon inscribed in the unit circle, where we have $|A-B| = |C-B|$ and $|A-C| = |A-D|$ by construction.

First show $|C - M| = \varphi$ and conclude $|A - C|^2 = 1 + \varphi^2 = 2 - \varphi$; use Exercise 28(b). Now you must show that the pentagon inscribed in the unit circle has the same side length. The two triangles in gray in the right figure are similar (why?). The proportions in the left of these triangles follow from Exercise 28(a). Now conclude from the right triangle: $(s/2)^2 = 1 - (\varphi + 1)^2/4 = (2 - \varphi)/4$.

Exercise 31 *Tetrahedral angle*

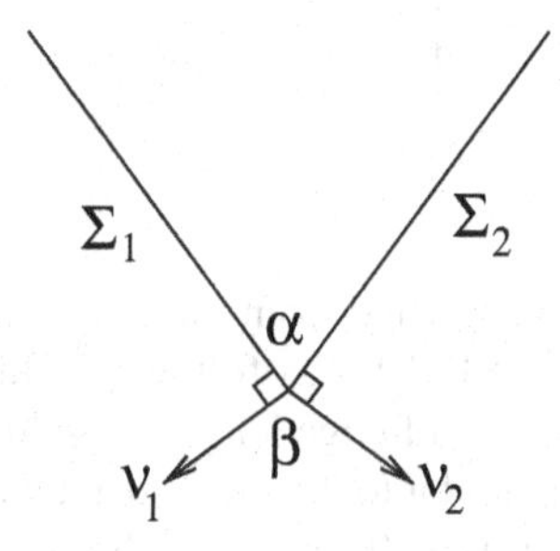

Calculate the edge angle α of the regular tetrahedron with the vertices $e_1, ..., e_4 \in \mathbb{R}^4$.

Note: A normal vector of the side Σ_1 with vertices e_2, e_3, e_4 is $v = e_1$, but this is not tangent to the hyperplane H through the points $e_1, ..., e_4$; for this we need to subtract from v its component in the direction of the vector $d = (1, 1, 1, 1)$ which is perpendicular to H. So the correct normal vector is $v_1 = e_1 - \frac{\langle e_1, d\rangle}{\langle d, d\rangle} d$.

Calculate the corresponding angle also for the n-dimensional regular simplex.

Exercise 32 *Construction of the dodecahedron*[9]

[9] For the construction of the icosahedron see Exercise 10.

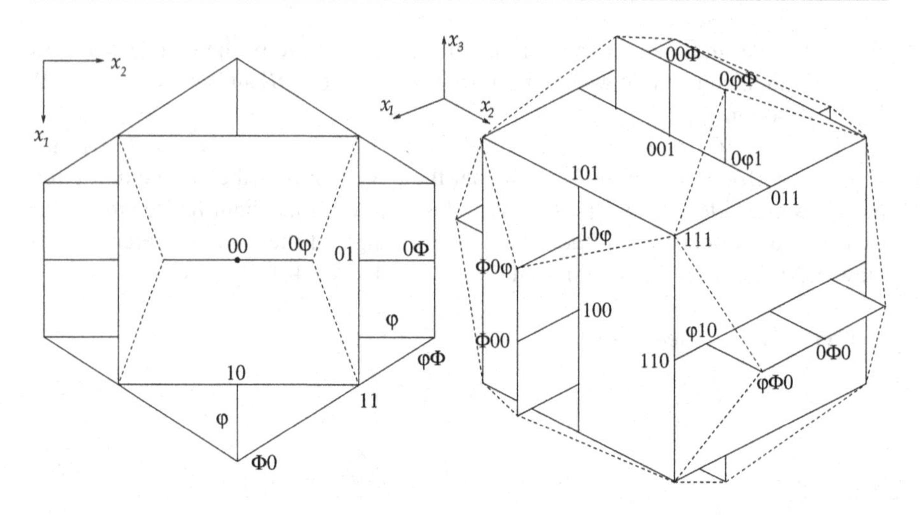

The dodecahedron can be constructed from the cube with vertices $(\epsilon_1, \epsilon_2, \epsilon_3)$ with $\epsilon_i \in \{\pm 1\}$ (shorthand: $(\pm 1, \pm 1, \pm 1)$) as follows. In the middle of each face one puts up side by side two squares of edge length φ (golden section < 1), both lying in a coordinate plane perpendicular to the face, such that all three coordinate planes occur (see right figure above). The 20 vertices of the dodecahedron are the 8 vertices of the cube together with the 12 free vertices of the planted double squares. The right figure shows the spatial situation, the left figure the intersection with the x_1x_2-plane (and at the same time the projection into this plane). The golden section is denoted by $\Phi > 1$ and $\varphi = 1/\Phi$. The pairs or triples of numbers (written without commas and parentheses, e.g. 11, $\Phi\varphi$ or $\Phi 0\varphi$) denote the coordinates of the points in the x_1x_2-plane or in $x_1x_2x_3$-space, respectively.
Convince yourself that the construction does indeed give the regular dodecahedron. Use the calculation rules of the golden section: $\Phi^2 = 1 + \Phi$, $\varphi = 1/\Phi = \Phi - 1$, $\varphi^2 = 1 - \varphi$. In detail, you should show:

(a) All 20 points, e.g., $(1, 1, 1)$ and. $(\varphi, \Phi, 0)$ lie on the sphere centered at the origin $(0, 0, 0)$ with radius $\sqrt{3}$.
(b) Each two adjacent pairs of points, e.g. $(\varphi, \Phi, 0)$, $(-\varphi, \Phi, 0)$ and $(\varphi, \Phi, 0)$, $(1, 1, 1)$ have distance 2φ.
(c) The points of the pentagons lie in a common plane: Consider the figure on the left and show that the points $(\Phi, 0)$, $(1, 1)$ and (φ, Φ) lie on a common straight line (i.e. the difference vectors are linearly dependent).
(d) The pentagons are regular. Show in particular that the ratio diagonal/edge equals Φ, cf. Exercise 28(a).

Exercise 33 *Vertices of the 24-cell*[10]
Consider the cube in $\mathbb{R}^n$ with the vertices $(\pm 1, \ldots, \pm 1)$ and the *cocube* (also called *cross-polytope* or *hyperoctahedron*) which is the convex hull of the vertices $\pm 2e_1$, ..., $\pm 2e_n$. We consider the convex hull K of the union of the two sets of vertices. The figure shows the case $n = 3$.

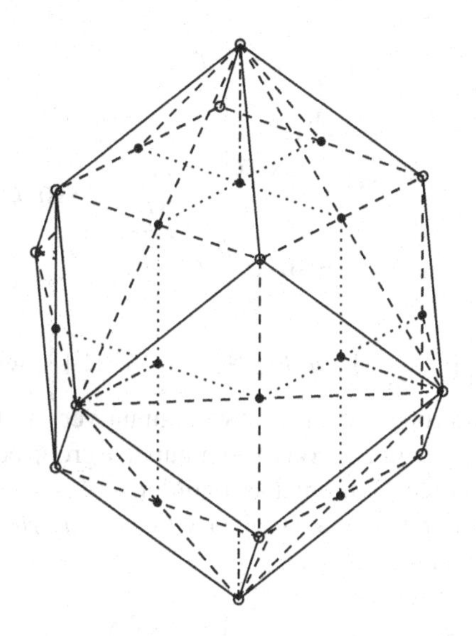

(a) Show that precisely for $n = 4$, the vertices of both the cube and the cocube have the same distance from the origin, i.e. they lie on a common sphere.
(b) Further show for $n = 4$ that the length of the cube edge is equal to the distance of adjacent cube and cocube vertices. Conclude that all edges of K have equal length.
(c) Show that, for example, the reflection along the perpendicular bisector on the line segment between (2, 0, 0, 0) and $(1, -1, -1, -1)$ keeps the union of the vertices of the cube and the cocube invariant (use symmetries!). So this reflection lies in the isometry group of K, which also still contains the (common) symmetry group of the cube and the cocube.
(d) Conclude that the isometry group of K acts transitively on the set of vertices of K.

[10] https://en.wikipedia.org/wiki/24-cell.

Exercise 34 *SO*(2) *and complex numbers, SU*(2) *and quaternions*

(a) The group *SO*(2) consists of the plane rotations $A = \begin{pmatrix} \cos\alpha & -\sin\alpha \\ \sin\alpha & \cos\alpha \end{pmatrix}$ for an angle $\alpha \in [0, 2\pi]$.

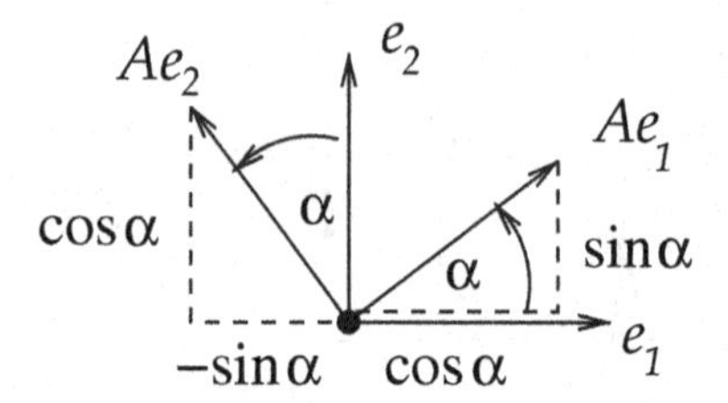

Thus $SO(2) = \left\{ \begin{pmatrix} a & -b \\ b & a \end{pmatrix};\ a, b \in \mathbb{R},\ a^2 + b^2 = 1 \right\}$, which can also be seen directly from the property that the two column vectors form an orthonormal basis with determinant one. This is a commutative group because plane rotations with center 0 commute. Omitting the condition $a^2+b^2 = 1$ we obtain a rotation composed with a homothety (also called a *direct similarity with center* 0). Show that the set of these matrices

$$\mathbb{C}' := \mathbb{R} \cdot SO(2) = \left\{ \begin{pmatrix} a & -b \\ b & a \end{pmatrix};\ a, b \in \mathbb{R} \right\}$$

forms a commutative subalgebra[11] of the matrix algebra $\mathbb{R}^{2\times 2}$. Moreover, $\mathbb{C}'$ is isomorphic (as an algebra) to the field $\mathbb{C}$ of complex numbers, where $1 = I = \left(\begin{smallmatrix} 1 & \\ & 1 \end{smallmatrix}\right)$ and $i = J = \left(\begin{smallmatrix} & -1 \\ 1 & \end{smallmatrix}\right)$.

(b) The *unitary group* is the complex analogue of the orthogonal group: $U(n) = \{A \in \mathbb{C}^{n\times n};\ A^*A = I\}$, where I is the unit matrix and $A^* = \bar{A}^t$ is the adjoint (= conjugate transposed) matrix: If $A = (a_{ij})$, then the coefficients b_{ij} of A^* are given by $b_{ij} = \overline{a_{ji}}$ (with the complex conjugation $\overline{x+iy} = x - iy$). The equation $A^*A = I$ says that the columns of A form an orthonormal basis with respect to the *Hermitian* scalar product $\langle v, w\rangle = v^*w = \sum \overline{v_i} w_i$ on $\mathbb{C}^n$. An important subgroup of $U(n)$ is the *special unitary group* $SU(n) = \{A \in$

[11] A vector space V with a bilinear map $\mu : V \times V \to V$ ("multiplication") is called an *algebra*. A *subalgebra* is a subspace $W \subset V$ which is invariant under multiplication: $\mu(W \times W) \subset W$.

$U(n);\ \det A = 1\}$. Show $SU(2) = \left\{ \begin{pmatrix} a & -\bar{b} \\ b & \bar{a} \end{pmatrix};\ a, b \in \mathbb{C},\ |a|^2 + |b|^2 = 1 \right\}$ and conclude that

$$\mathbb{H} := \mathbb{R} \cdot SU(2) = \left\{ \begin{pmatrix} a & -\bar{b} \\ b & \bar{a} \end{pmatrix};\ a, b \in \mathbb{C} \right\}$$

is a $\mathbb{R}$-subalgebra of $\mathbb{C}^{2\times 2}$ and all nonzero elements of $\mathbb{H}$ are invertible; $\mathbb{H}$ is therefore a skew field.

Exercise 35 *Focal points of ellipses and hyperbolas*

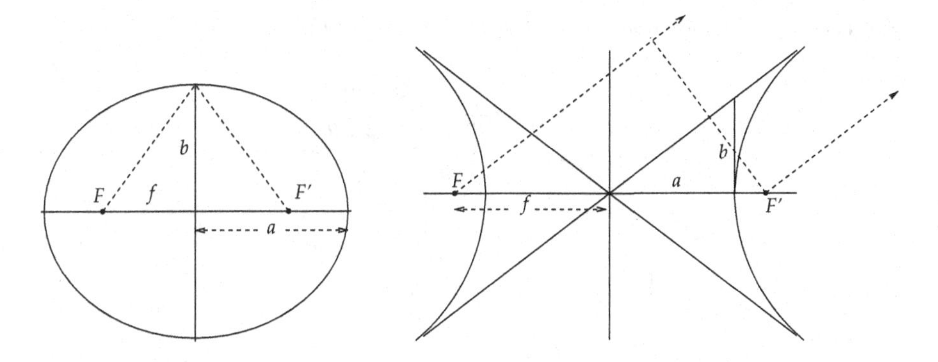

Given an ellipse and a hyperbola with semi-axes a and b. Let f denote the distance of the focal points from the center. Show:

(a) $f^2 = a^2 - b^2$ for the ellipse,
(b) $f^2 = a^2 + b^2$ for the hyperbola.[12]

Exercise 36 *Directrix*

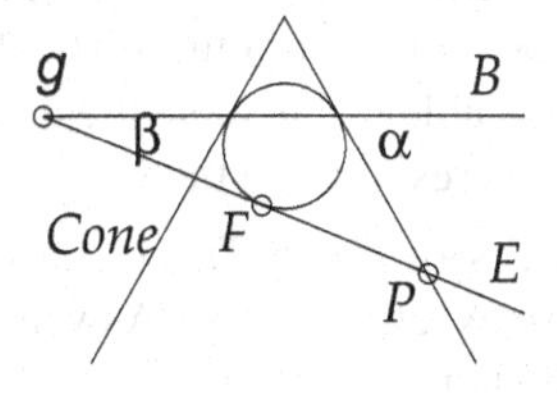

Show that there is a directrix not only for parabolas but also for ellipses and hyperbolas: It is the line g where the plane E of the conic section intersects the plane B of the circle where the Dandelin sphere touches the cone. However, the

[12] Use similarity of right triangles with a common angle.

distances of a conic section point P to the focal point F and to the directrix g are no longer equal, but they have a constant ratio: $|P, F|/|P, g| = const$ (independent of P).

Hints: One uses again that the distance $|P, F|$ equals the length of the segment of the generatrix between its intersection with B and P. The angles α and β, which separate the horizontal plane from the generatrix of the cone (α) and the plane E of the conic section (β) are independent of P.

8.4 Differential Geometry (Chap. 5)

Exercise 37 *Circumference of the Earth according to Eratosthenes*

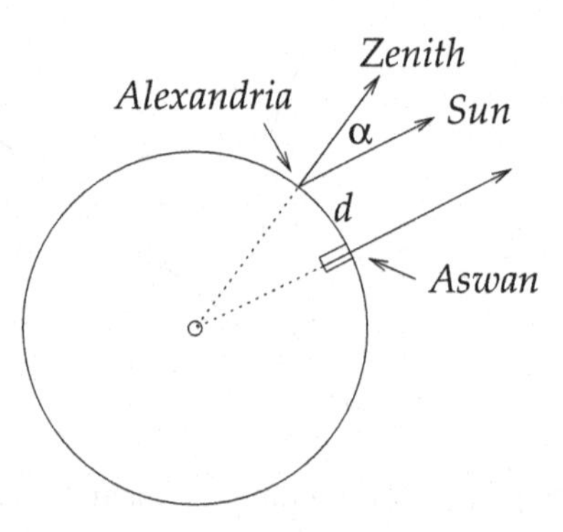

At the time of *Eratosthenes of Cyrene*, who lived in the ancient scientific center of Alexandria in Egypt about 275–196 BC, people were aware of a deep well at Aswan (then called Syene) into which the sun shone exactly vertically at noon on June 21 of each year, so that the otherwise dark surface of the water flashed brightly. At the same hour, Eratosthenes determined the position of the sun in Alexandria which is about 800 km north of Aswan. Measuring the length of the shadow of a vertical rod or plummet, he arrived at a deviation from the vertical of about $\alpha = 7.2°$ which is 1/50 of 360°. From this he was able to calculate the 40,000 km circumference of the Earth. How did he arrive at this result? Use the figure above.

There were, of course, sources of error: The distance was only known approximately, and Alexandria is not exactly north of Aswan, but 3° further east. Can you estimate the magnitude of the latter error?

Exercise 38 *Surfaces of revolution*
Let $f : (a, b) \to (0, \infty)$ be a C^2-function. Show that the mapping $\varphi : (a, b) \times \mathbb{R} \to \mathbb{R}^3$, $\varphi(s, t) = (f(s) \cos t, f(s) \sin t, s)$ is an immersion. Determine the coefficients of the fundamental forms g and h (i.e. $g_{ss}, g_{st}, g_{tt}, h_{ss}, h_{st}, h_{tt}$), the principal curvatures, and the Gaussian and mean curvatures. Give a necessary and sufficient condition for f such that φ is a minimal surface ($H = 0$).

Exercise 39 *Orthogonal families of quadrics*

http://mathworld.wolfram.com/ConfocalQuadrics.html

Let $a_1 < a_2 < \ldots < a_n$ be real numbers. For $j = 1, \ldots, n$ we set $I_j = (a_{j-1}, a_j)$ with $a_0 := -\infty$. For each $u \in \bigcup_j I_j$ we consider on $\mathbb{R}^n$ the quadratic polynomial $q^u(x) = \sum_{i=1}^n \frac{x_i^2}{a_i - u}$ and the corresponding quadric $Q_u = \{x;\ q^u(x) = 1\}$. Let $\nabla q^u := (\frac{\partial q^u}{\partial x_1}, \ldots, \frac{\partial q^u}{\partial x_n})^T$ denote the *gradient* of q^u. Show that $\langle \nabla q^u, \nabla q^v \rangle = 4\,\frac{q^u - q^v}{u - v}$. Conclude that the n sets of quadrics $(Q_u)_{u \in I_j}$ for $j = 1, \ldots, n$ form an orthogonal hypersurface system; recall that the gradient of q^u at $x \in Q_u$ is perpendicular to the tangent space $T_x Q_u$. Sketch the situation for $n = 2$.

8.5 Conformal Geometry (Chap. 6)

Exercise 40 *Mercator map*
The Mercator map[13] is an angle-preserving mapping $\mu : \mathbb{S}^2 \setminus \{\pm N\} \to \mathbb{R}^2 = \mathbb{C}$ (with $N = e_3 = (0, 0, 1)$) with the following properties:

(1) Latitude circles go onto parallels of (part of) the real axis $\mathbb{R}$, where the longitude φ equals the arc length along the image[14]; "east" is right,
(2) Meridians go onto parallels of the imaginary axis $i\mathbb{R}$, where the image arc length is strictly monotonic increasing with the latitude θ; "north" is up,
(3) The equator is mapped isometrically onto $[-\pi, \pi] \subset \mathbb{R}$.

(Note that we have previously replaced the globe with circumference 40,000 km to the unit sphere $\mathbb{S}^2$ with circumference 2π)

[13] See e.g. https://kartenprojektionen.de/license/mercator-84:flat-ssw, Gerhard Mercator (Cartographer), 1512 (Rupelmonde, Belgium)–1594 (Duisburg).
[14] We disregard the periodicity of the latitude φ.

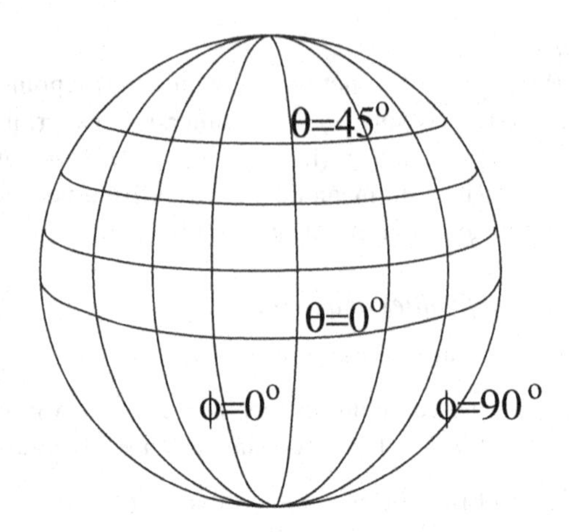

(a) Show that the conformal factor $\lambda(x) = |d\mu_x(v)|/|v|$ at each point $x \in \mathbb{S}^2$ has the value $1/\cos\theta$ where $\theta(x)$ is the latitude of $x \in \mathbb{S}^2$ (that means $\theta = 0°$ at the equator, $\theta = 90°$ at the north pole).

(b) Conclude that the Mercator map is uniquely determined by (1),(2),(3): If $\varphi(x) \in [-\pi, \pi]$ is the longitude of x and conversely $x(\varphi, \theta) \in \mathbb{S}^2$ is the point with longitude φ and latitude θ, then $\mu(x(\varphi, \theta)) = \varphi + i \int_0^\theta (1/\cos(t))dt \in \mathbb{C}$.

(c) The inverse of μ is the mapping $\alpha : z \mapsto \Phi(e^{i\bar{z}})$ where $\Phi : \mathbb{C} = \mathbb{R}^2 \to \mathbb{S}^2 \setminus \{e_3\}$ is the stereographic projection.

Hint: Don't try to compute! Just verify that the inverse of α satisfies the properties (1)–(3). Note that the interior map of the composition α is antiholomorphic and in particular conformal, and it transforms $z = s + it$ into $e^{i\bar{z}} = e^t e^{is}$; the coordinate lines $\mathbb{R} + it$ and $s + i\mathbb{R}$ thus go to the circle around 0 with radius e^t and to the radial ray which has angle s with the positive real axis.

Exercise 41 *Stereographic projection and inversion*

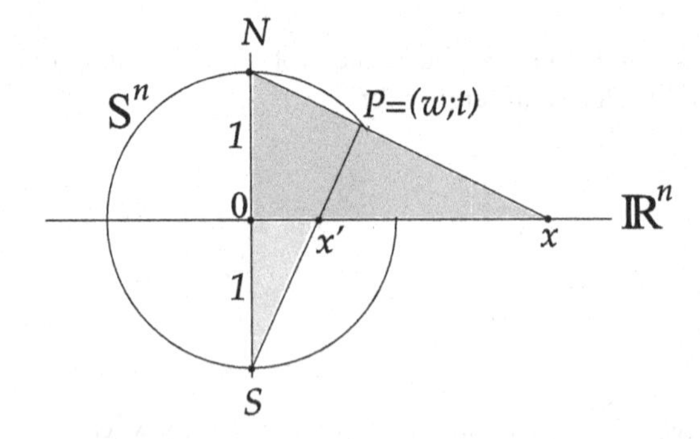

The analytical formulas for the inverse stereographic projection $\Psi_+ : \mathbb{S}^n \setminus \{N\} \to \mathbb{R}^n$ from the "north pole" $N = e_{n+1}$ and their inverse Φ_+ are given in (6.8). From this, develop the corresponding formulas for the inverse stereographic projection from the "south pole", $\Psi_- : \mathbb{S}^n \setminus \{-N\} \to \mathbb{R}^n$. Conclude (a) analytically and (b) geometrically (note and justify the similarity of the two colored triangles in the figure) that $\Psi_- \circ \Phi_+ =: F$ is the inversion along the unit sphere in $\mathbb{R}^n$. (What is then $\Psi_+ \circ \Phi_-$?) Thus conformality and sphericality of the inversion follow also from the corresponding properties for the stereographic projection (which we had proven geometrically).

Exercise 42 *Chord theorem and inverter*

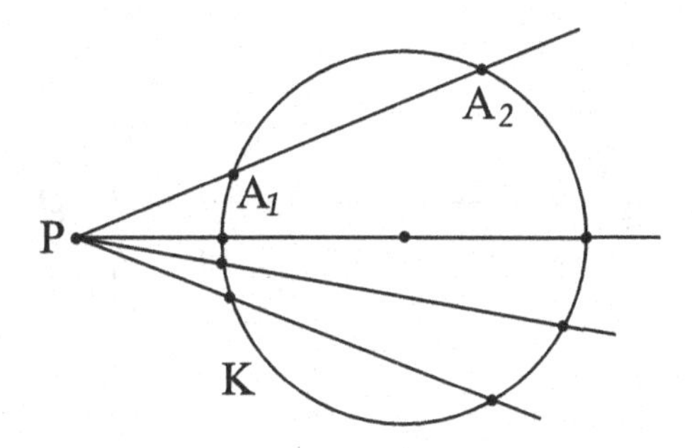

(a) Let K be a circle in the plane and P a point outside. Consider any straight line g through P that intersects K at two points; denote the intersection points A_1 and A_2. Show that the product of the distances $|P - A_1||P - A_2|$ is independent of g (*chord theorem*).
Note: The center of K *is* 0. *Parametrize* g *by* $g(t) = P + tv$ *with* $|v| = 1$ *and determine the coefficients of the quadratic equation* $|g(t)|^2 = r^2$. *The product of the two solutions* t_1, t_2 *of a quadratic equation* $t^2 + at + b = 0$ *is* b (why?).

(b) Using (a), show that the rod construction shown in the figure is an *inverter*, a mechanical device that accomplishes inversion along a circle, i.e., 0, x, Fx are collinear and $|x||Fx| = s^2 - r^2$ (compare (6.6)).

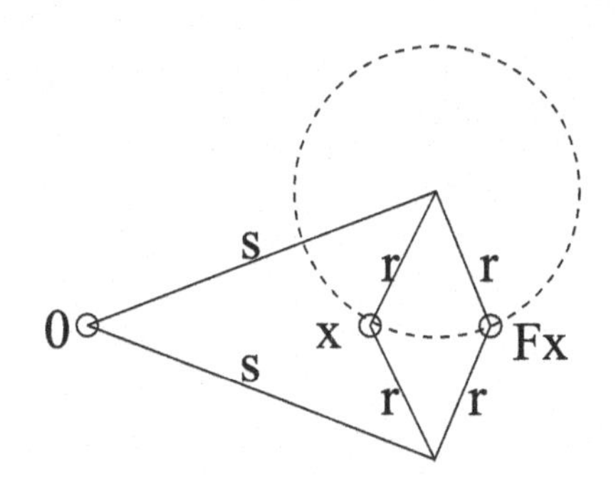

Exercise 43 *Composition of circular inversions*
First show that the composition of two inversions along concentric circles in the plane $\mathbb{R}^2$ (you can use the origin O as the common center) is a homothety. This fixes the common center and leaves invariant the radial rays emanating from the center. Now consider two circles k_1, k_2 such that k_1 is still contained in k_2, but no longer concentric (see figure below), and study the composition of the two associated inversions. Show that this composition also has a fixed point and that a family of circles through this point remains invariant!

Note: There is a circle k which meets perpendicularly the circles k_1 and k_2 and also the straight line g connecting the two centers. Let A, B be the intersections of k with g. The inversion along an arbitrary circle with center at B transforms k_1 and k_2 into concentric circles again—why?

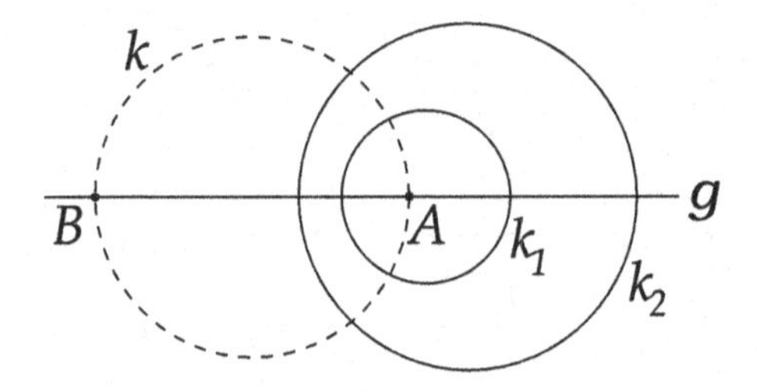

8.6 Spherical and Hyperbolic Geometry (Chap. 7)

Exercise 44 *Spherical triangles*
In the Euclidean plane, the sum of interior angles for any triangle is known to be $\pi = 180°$ (see the first figure in Chap. 1). For triangles on the sphere of radius one, whose edges are great circle arcs, it is quite different. The sum of the interior angles is always greater than π, and the excess $\alpha + \beta + \gamma - \pi$ is equal to the *area* F of the triangle. Show this with the help of the following figure:

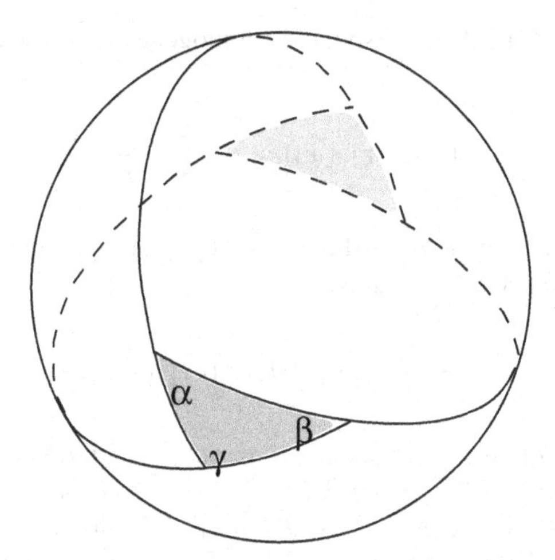

Hints: We consider the three spherical biangles[15] *with angles $\pi-\alpha, \pi-\beta, \pi-\gamma$ adjacent to the given triangle (shaded dark in the figure). The great circle arcs bounding a biangle reunite at the opposite ("antipodal") point. Thus the whole surface of the sphere is decomposed into: the triangle, the three biangles, and the image of the triangle under the antipodal map $-I$ (shaded light in the figure). The area of a biangle with angle δ is the fraction $\delta/(2\pi)$ of the total unit sphere area (why?). But the area of the sphere is 4π according to Archimedes,*[16] *so the biangle with angle δ has the area 2δ. From this and from the above decomposition of the sphere the assertion follows.*

Exercise 45 *Hyperbolic distance and cross-ratio*
For $x, y \in \mathbb{R}^{n+1}$ denote by

$$\langle x, y\rangle_- = x_1y_1 + \cdots + x_ny_n - x_{n+1}y_{n+1}$$

the indefinite scalar product of Lorentz and Minkowski. Consider two linearly independent vectors

$$v, w \in H := \{x \in \mathbb{R}^{n+1};\ \langle x, x\rangle_- = -1,\ x_{n+1} > 0\}.$$

These span a plane $E = \mathbb{R}v + \mathbb{R}w$. The intersection of E with the light cone $C = \{x \in \mathbb{R}^{n+1};\ \langle x, x\rangle_- = 0\}$ consists of two straight lines $l_1 = \mathbb{R}n_1$ and $l_2 = \mathbb{R}n_2$ for two linearly independent vectors $n_1, n_2 \in C \cap E$. For each $x \in \mathbb{R}^{n+1}_*$

---

[15] *A spherical biangle (lune, digon) is a domain on the two-dimensional sphere which is bounded by two half great circles.*

[16] See e.g. "Sternstunden der Mathematik" [12], p. 27.

let $[x] = \mathbb{R}_* x \in \mathbb{RP}^n$ be the corresponding homogeneous vector, and $(\, , \, ; \, , \,)$ denote the cross-ratio:

$$([x], [y]; [z], [w]) = \frac{z-x}{z-y} : \frac{w-x}{w-y},$$

where the representatives x, y, z, w lie on an (affine) straight line in $\mathbb{R}^{n+1}$. Show for the hyperbolic distance a between v and w:

$$a = \frac{1}{2} |\log |([v], [w]; [n_1], [n_2])||.$$

Note: Both sides of this equation are invariant under all projective mappings preserving the sphere $\mathbb{S}^{n-1} = [C]$ *(why?). Therefore, we can assume without restriction* $v = e_{n+1}$ *and* $w = (\cosh a)e_{n+1} + (\sinh a)e_1$ *(why?). Choose the representatives on the straight line* $g = e_{n+1} + \mathbb{R}e_1$ *(see figure below). In particular, the representative of* $[w]$ *on this straight line is the vector* $\tilde{w} = e_{n+1} + t \cdot e_1$ *with* $t = \tanh a = \frac{\sinh a}{\cosh a}$*. Now you can calculate the cross-ratio; recall* $\cosh a = \frac{1}{2}(e^a + e^{-a})$ *and* $\sinh a = \frac{1}{2}(e^a - e^{-a})$.

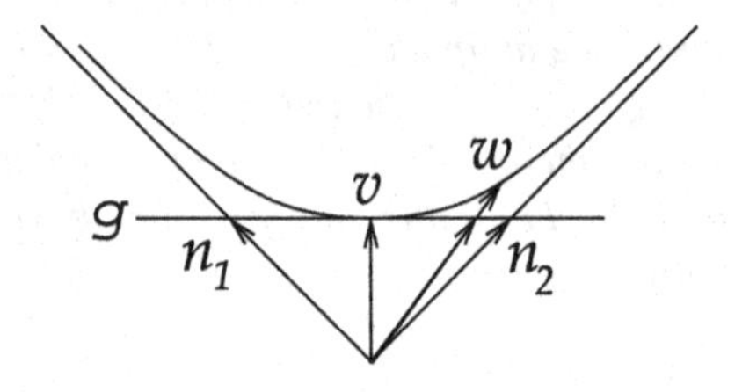

Solutions

9

Abstract

Here, the exercise problems set in the previous chapter are largely solved. We recommend our readers not to look at the solutions until they have worked intensively on the exercises themselves.

1. (a) Left figure: direction of the sun slightly to the right of the direction of the moon (at new moon the direction of the sun and the direction of the moon would be almost identical), middle figure: direction of the sun further to the right at right angle to the direction of the moon, right figure: almost straight line sun - observer - moon.
   (b) Middle figure: afternoon or early evening before sunset; since the sun is 90° to the right of the moon and both run approximately the same path from left to right, the sun is ahead of the moon, i.e. further to the west. Therefore it cannot be morning. Left figure: day, right figure: night.
2. (a) $x = w_e x \Rightarrow x \sim x$; $x \sim y \Rightarrow \exists_{g \in G}\ y = w_g x \Rightarrow x = w_{g^{-1}} y \Rightarrow y \sim x$; $x \sim y,\ y \sim z \Rightarrow \exists_{g,h \in G}\ y = w_g x,\ z = w_h y \Rightarrow z = w_h y = w_h w_g x = w_{hg} x \Rightarrow x \sim z$.
   (b) $e \in G_x$ because $w_e x = x$, $g \in G_x \Rightarrow w_g x = x \Rightarrow x = w_{g^{-1}} x \Rightarrow g^{-1} \in G_x$, $g, h \in G_x \Rightarrow w_g x = x = w_h x \Rightarrow w_{gh} x = w_g w_h x = w_g x = x \Rightarrow gh \in G_x$.
   (c) Well-definedness: $gG_x = hG_x \Rightarrow k := g^{-1}h \in G_x \Rightarrow h = gk$ with $k \in G_x \Rightarrow w_h x = w_{gk} x = w_g w_k x = w_g x$. Injectivity: $w_g x = w_h x \Rightarrow x = w_{g^{-1}} w_h x = w_{g^{-1}h} x \Rightarrow k := g^{-1}h \in G_x \Rightarrow h = gk,\ k \in G_x \Rightarrow hG_x = gkG_x = gG_x$. Surjectivity: $y \in Gx \Rightarrow \exists_{g \in G}\ y = w_g x = w^x(gG_x) \Rightarrow y \in \text{image}\ w^x$.
3. A vertex x can be rotated into any vertex y. If we have two such rotations g, h with $gx = hx = y$, then $h^{-1}gx = x$; the rotation $k := h^{-1}g$ is therefore a rotation of the cube that leaves the vertex x fixed. Such rotation permutes the

J.-H. Eschenburg, *Geometry – Intuition and Concepts*,
https://doi.org/10.1007/978-3-658-38640-5_9

three edges that meet in x, and there are three even permutations (determinant one): They correspond to the identity and to the rotations by 120 and 240° in the plane perpendicular to x. Therefore there are exactly three rotations of the cube that take x into y. There are 8 vertices y, hence we have $3 \cdot 8 = 24$ rotations of the cube.

The cube group G acts transitively on the set of vertices X, so $Gx = X$ for each $x \in X$. According to 2(c), $|G/G_x| = |Gx| = |X| = 8$ (by $|S|$ we denote the number of elements, the *cardinality* of a set S). Moreover, we have seen that three rotations leave the vertex x fixed, i.e. $|G_x| = 3$. Since all cosets gG_x have the same number of elements, viz. $|gG_x| = |G_x| = 3$, we see again $|G| = |G/G_x||G_x| = |X||G_x| = 8 \cdot 3 = 24$.

4. Since homotheties preserve straight lines through the center and transform each straight line into a parallel, there is a homothety $S_\lambda : x_1 \mapsto x_2,\ y_2 \mapsto y_1$ and another $S_\mu : x_2 \mapsto x_3,\ y_3 \mapsto y_2$. Then $x_3 = S_\mu x_2 = S_\mu S_\lambda x_1 = S_{\mu\lambda} x_1$ and $y_1 = S_\lambda y_2 = S_\lambda S_\mu y_3 = S_{\lambda\mu} y_3$. The straight lines $x_1 y_3$ and $x_3 y_1$ are parallel if and only if there is a homothety S_α with $S_\alpha x_1 = x_3$ and $S_\alpha y_3 = y_1$, i.e. $S_\alpha(x_1) = S_{\mu\lambda}(x_1)$ and $S_\alpha(y_3) = S_{\lambda\mu}(y_3)$. Since a homothety is determined by its value on any point outside the center o, this is equivalent to $S_{\lambda\mu} = S_\alpha = S_{\mu\lambda}$, i.e. to $\lambda\mu = \mu\lambda$.

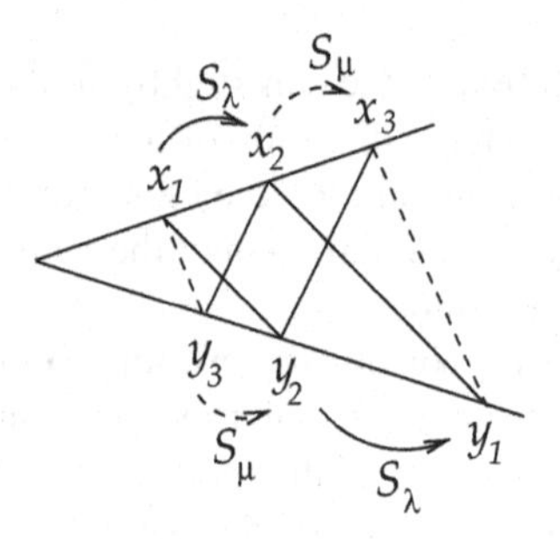

5. The two triangles are x, y, z and $\lambda x, \lambda y, \lambda z$. According to the geometric definition of S_λ two pairs of sides are already parallel: $x \vee y \parallel \lambda x \vee \lambda y$ and $x \vee z \parallel \lambda x \vee \lambda z$ (where $A \vee B = AB$ denotes the straight line through two points A, B). Since S_λ preserves directions, the third pair of lines is parallel too: $y \vee z \parallel \lambda y \vee \lambda z$.
6. We need to show that Aff(X) is a subgroup of the group Bij(X) of all bijective mappings on X. In fact, $F, G \in$ Aff(X) with $Fx = Ax + a$, $Gx = Bx + b$ implies $FGx = A(Bx + b) + a = ABx + Ab + a$, so $FG \in$ Aff(X). The neutral element of Bij(X) is id_X; this is in Aff(X), for $\mathrm{id}_X x = Ix + 0$, where I is the unit matrix on X. Finally, $FG = \mathrm{id}_X$ if $ABx = x$ and $Ab + a = 0$, thus $B = A^{-1}$ and $b = -A^{-1}a$. Hence the inverse of F is the affine mapping $Gx = A^{-1}x - A^{-1}a$.

Each pair $(A, a) \in GL(X) \times X$ determines the affine mapping $F : x \mapsto Ax + a$, and vice versa an affine mapping F determines the pair (A, a) because

$a = F(0)$ and $A : x \mapsto F(x) - F(0)$. Therefore we have a bijective mapping $\Phi : GL(X) \times X \to \mathrm{Aff}(X)$. The composition FG corresponds to the pair $(AB, a + Ab)$; we can take this to be the definition of a group structure on $GL(X) \times X$ as a *semidirect product*, which makes Φ a group isomorphism.

7. We have $ns_i + a_i = \sum_j a_j = (n+1)s$, that is $s = \frac{n}{n+1}s_i + \frac{1}{n+1}a_i = \lambda a_i + (1-\lambda)s_i$ with $\lambda = \frac{1}{n+1}$, therefore a_i, s, s_i are collinear. Furthermore $a_i - s = (1-\lambda)(a_i - s_i)$ and $s_i - s = \lambda(s_i - a_i)$, therefore $(a_i - s)/(s_i - s) = \frac{\lambda - 1}{\lambda} = 1 - \frac{1}{\lambda} = 1 - (n+1) = -n$.
8. (a) The median divides the side into two equal parts. If we take these as the base sides of the two partial triangles, we see that the areas are equal, because the height (distance from the base side to the opposite point) is also the same.
   (b) The statement is false: The median is subdivided by the centroid (center of gravity) by the ratio $\frac{2}{3} : \frac{1}{3}$. A parallel to the base side through the centroid divides the triangle into a similar triangle (same angles) whose sides are smaller by the factor 2/3, and a trapezium. The smaller triangle is obtained from the larger one by a homothety with scaling factor 2/3; its area is therefore $(2/3)^2$ times the total area. But $(2/3)^2 = 4/9 < 1/2$.

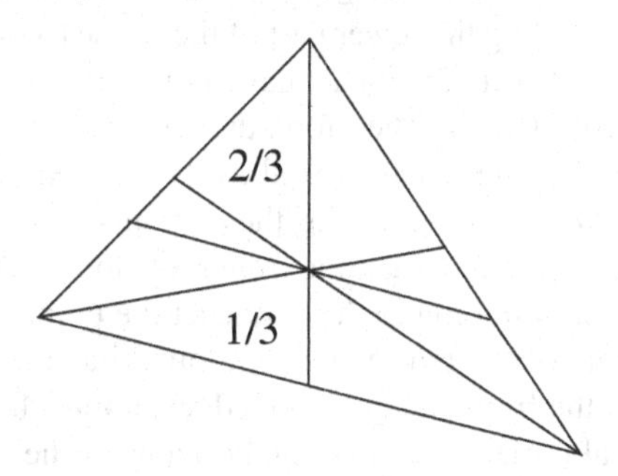

9. (a) A median partitions the corresponding side of Δ into two equal parts; thus, according to the ray theorem each straight line section parallel thereto within Δ is divided into two equal parts, in particular the side of δ which the median intersects. The medians of Δ are therefore also medians of δ and their common point of intersection is the centroid of both δ and Δ.
   (b) The medians in Δ connect the vertices of δ and Δ and are divided by the centroid s in the ratio $\frac{1}{3} : \frac{2}{3}$ (cf. Exercise 7), i.e. $1 : 2$. Therefore, the vertices of Δ arise from those of δ by a homothety S with center s and scaling factor -2.
   (c) An altitude on one side of δ intersects the opposite side of Δ perpendicularly and in the middle, so it is a perpendicular bisector. Therefore, the intersection point h of the altitudes of δ (the orthocenter) is also the intersection point M of the perpendicular bisectors (the center of the circumcircle) of Δ.
   (d) The homothety S of (b) maps δ to Δ and therefore also the center of the circumcircle m of δ to the circumcircle center $M = h$ of Δ. The common center of gravity (centroid) s remains fixed. We choose s as the origin: $s = 0$.

Then $h = S(m)$, thus $h = -2m$, and h, s, m lie on a common straight line with $(m - s)/(h - s) = -2$.

10. (b) The three subdivision points on the octahedron face $e_1e_2e_3$ are $p_1 = \varphi e_1 + \varphi^2 e_2 = (\varphi, \varphi^2, 0)$, $p_2 = \varphi e_2 + \varphi^2 e_3 = (0, \varphi, \varphi^2)$, $p_3 = \varphi e_3 + \varphi^2 e_1 = (\varphi^2, 0, \varphi)$. Then $p_1 - p_2 = \varphi(1, \varphi - 1, -\varphi)$. By $1 - \varphi = \varphi^2$ follows $|p_1 - p_2|^2 = \varphi^2(1 + (\varphi - 1)^2 + \varphi^2) = \varphi^2(2(1 - \varphi) + 2\varphi^2) = \varphi^2 \cdot 4\varphi^2$ and thus $|p_1 - p_2| = 2\varphi^2$. A neighbor point to p_1 on an adjacent octahedron face is $p_4 = \varphi e_1 - \varphi^2 e_2 = (\varphi, -\varphi^2, 0)$ and $p_1 - p_4 = (0, 2\varphi^2, 0)$ obviously also has the length $2\varphi^2$.
11. The vertical edges are shown as vertical straight lines. The lower and upper edges of the gable front (up to the base of the roof) must both run obliquely upwards and meet at a point P which is higher than the top of the roof, so that the house is seen from above (bird's eye view). The horizon of the horizontal planes (e.g. floor level) is the horizontal straight line h through P, the horizon of the gable front plane is the vertical line v through P. The alignment lines of the right side wall (extensions of the top and bottom edges) must also meet on the horizontal horizon h, at a point $Q \in h$. The horizon of the side wall plane is the vertical straight line w through Q. Now we can easily complete the drawing to the cuboid representing the lower part of the house (without the roof). Since the gable front is square, its diagonals have in reality 45° slope and are therefore parallel to the roof slopes. Therefore, draw the two diagonals of the gable end (determined by their vertices) as well as their points of intersection R, S with the vertical horizon v. Connecting these points to the top two points of the gable front quadrilateral, we get the front roof slopes. The rear ones are parallel to the front ones; in the figure they connect the points at infinity R and S with the upper corners of the rear gable front quadrilateral. The ridge connects the intersections of the front and rear roof edges; it must be directed to the point Q of the horizontal horizon. Finally, the horizons of the two roof planes must be drawn; these are the straight lines RQ and SQ.
12. The center of projection is $Z = (0, 1, 1)$. Let the original point be $P = (x, y, 0)$. The projection line $p = PZ$ is thus parametrized by $p(t) = P + t(Z - P) = (x, y, 0) + t(-x, 1 - y, 1) = (x - tx, y + t(1 - y), t)$. The image point is the intersection of p with the xz-plane, in which the y-coordinate vanishes. Accordingly t is to be chosen in such a way that $y + t(1 - y) = 0$, thus $t = \frac{y}{y-1}$ and $t - 1 = \frac{1}{y-1}$. Therefore the image point $P' = (x', 0, z') = p(\frac{y}{y-1})$ has the coordinates $x' = x(1 - t) = -\frac{x}{y-1}$ and $z' = t = \frac{y}{y-1}$. Conversely: $(y - 1)z' = y$ $\Rightarrow y(z' - 1) = z' \Rightarrow y = \frac{z'}{z'-1}$, and $x = -x'(y - 1) = -\frac{x'}{z'-1}$. If now a set of parallel lines $ax + by = s$ is given ($a, b \in \mathbb{R}$ fixed, $s \in \mathbb{R}$ the variable parameter), we obtain the equation of the image straight lines by substituting $x = -\frac{x'}{z'-1}$ and $y = \frac{z'}{z'-1}$, thus $-ax' + bz' = s(z' - 1)$. An intersection point $(x', 0, z')$ of all image lines satisfies this equation for all s simultaneously; this is only possible if the s-terms vanish, which happens precisely for $z' = 1$; we then obtain $-ax' + b = 0$ and thus $x' = \frac{b}{a}$. So the image lines intersect at

the point $(\frac{b}{a}, 0, 1)$ (*vanishing point*). The horizon, which contains all vanishing points, is the straight line $\{(x', 0, 1);\ x' \in \mathbb{R}\}$.

13. The horizon of all "horizontal" planes is the straight line EF, the viewpoint of the observer is H. The points $E, 0, F$ deviate from the viewing direction by 45°. The table and the chair are trickily placed in such a way that the edges of the chair (partially hidden, but recognizable from the position of the chair legs) are parallel to the diagonals of the tabletop, for their vanishing lines intersect at the horizon points E and F. We may assume that the chair edges intersect (in reality) at right angles. Thus the diagonals of the tabletop also intersect at right angles. By premise, the tabletop is a rectangle. Since its diagonals intersect at right angles, this rectangle is a square.

    We can see the true angles if we rotate the horizontal plane by 90° around the horizon EF (passing through H and G) into the image plane, which is rendered undistorted (on any plane parallel to the image plane, the central projection is, after all, a homothety).

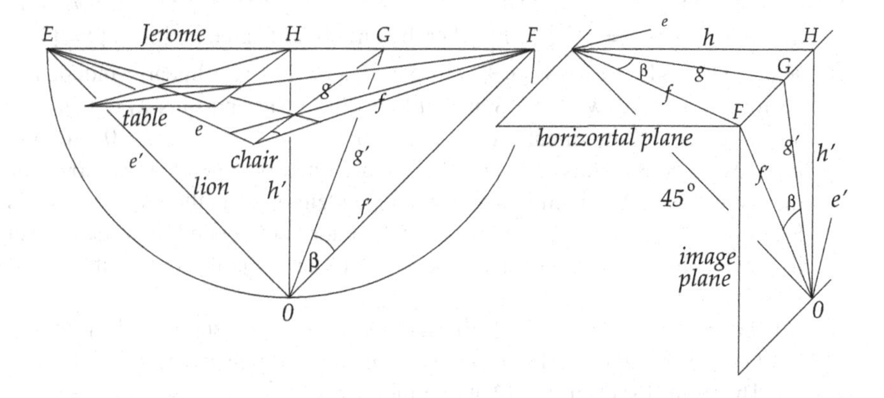

The edges e and f of the chair become the straight lines e' and f' which meet at point 0 at the correct 90°-angle, and also the rotated table edge h' and the rotated diagonal e' meet at the correct angle 45°. Likewise, the rotated versions of the chair edge and diagonal, f' and g', enclose the "true" angle β. The angle has not been changed by the rotation of the plane, and in the vertical image plane we can see it undistorted, see the right part of the figure.

14. (a) A is wrong. It is true that a projective mapping F of $\mathbb{RP}^2$ is obtained from a linear mapping A on $\mathbb{R}^3$ with $F[x] = [Ax]$, and indeed A is given by the images of three basis vectors, the vectors Ab_1, Ab_2, Ab_3, but the three image points $[a_i] = F[b_i] = [Ab_i] \in \mathbb{RP}^2$ determine Ab_i only up to scalar multiples. After fixing a_i we only know that $Ab_i = \lambda_i a_i$ with unknown scalars $\lambda_i \in \mathbb{R}$. Thus $F = [A]$ is not yet determined. It is certainly true that a plane in (affine or projective) space is determined by three points in general position, but we are not talking about the position of a plane in space, but about a projective mapping of the plane onto itself which is a different subject.

C is only partially right; indeed, the image of a square under a perspective mapping (photo) is always a convex quadrilateral. But, after all, we are taking a photograph in a particular direction, so we are only imaging a half-plane or half-space. If we use central projection to image the whole plane (i.e. including the region "behind the photographer"), we can easily find a square whose image is no longer convex, see (b). By definition, convex figures lie on one side of their supporting lines, but the term "on one side" makes no sense in projective plane, since this is not orientable (it contains a Möbius strip, see end of Sect. 3.3).

(b) The given vertices of the square are $[b_3]$, $[b_1]$, $[b_2]$ and $[b_1+b_2-b_3]$, the image points $[-b_3] = [b_3] = [0, 0, 1]$, $[b_1] = [1, 0, 1]$, $[b_2] = [0, 1, 1]$ and $[b_1+b_2+b_3] = [1, 1, 3] = [\frac{1}{3}, \frac{1}{3}, 1]$. All four image points lie in the affine plane $\mathbb{A}^2 = \{[x, y, 1];\ x, y \in \mathbb{R}\} \cong \mathbb{R}^2$ and form there the quadrilateral with the vertices $(0, 0)$, $(1, 0)$, $(0, 1)$, $(\frac{1}{3}, \frac{1}{3})$; this is not convex.

(c) Compare Exercise 20.

15. $F = [A]$ for $A = \left(\begin{smallmatrix} a & b \\ c & d \end{smallmatrix}\right)$ with $ad - bc = \det A \neq 0$. So $F[x, 1] = \left[\left(\begin{smallmatrix} a & b \\ c & d \end{smallmatrix}\right)\left(\begin{smallmatrix} x \\ 1 \end{smallmatrix}\right)\right] = [ax + b, cx + d] = \left[\frac{ax+b}{cx+d}, 1\right]$. Thereby one should agree on $[\infty, 1] := [1, 0]$.

16. $F = [S] = \mathrm{id} \iff \forall_{v \in V} \exists_{\lambda_v \in \mathbb{K}^*}\ Sv = \lambda_v v$. For two linearly independent vectors $v, w \in V$ we have $S(v + w) = Sv + Sw$ and thus $\lambda_v v + \lambda_w w = \lambda_{v+w}(v + w)$. Consequently $(\lambda_v - \lambda_{v+w})v + (\lambda_w - \lambda_{v+w})w = 0$ and hence $\lambda_v = \lambda_{v+w} = \lambda_w$, thus $\lambda_v = \lambda_w$. If v, w are linearly dependent, we look for a third vector $u \in V$ which is linearly independent of v, w; then $\lambda_v = \lambda_u = \lambda_w$. So we get $Sv = \lambda v$ for some $\lambda \in \mathbb{K}^*$ which does not depend on v. Conversely, it is clear that $F = [\lambda I]$ for any $\lambda \in \mathbb{K}^*$ (where I is the unit matrix) is the identity on P_V.

17. (a) We have $x^\sigma + y^\sigma = (x + y)^\sigma$ and $(\lambda x)^\sigma = \lambda^\sigma x^\sigma$ since $(x_i)^\sigma + (y_i)^\sigma = (x_i + y_i)^\sigma$ and $(\lambda x_i)^\sigma = \lambda^\sigma (x_i)^\sigma$ for all components x_i, $i = 1, \ldots, n$. Therefore for each $\sigma \in \mathrm{Aut}(\mathbb{K})$ the mapping $\hat{\sigma} : \mathbb{K}^n \to \mathbb{K}^n$, $x \mapsto x^\sigma$ is semilinear. Since $\hat{\sigma}\hat{\tau} = \widehat{\sigma\tau}$ and $\widehat{\mathrm{id}} = \mathrm{id}_{\mathbb{K}^n}$, the map $\sigma \mapsto \hat{\sigma} : \mathrm{Aut}(\mathbb{K}) \to \Gamma L(\mathbb{K}^n)$ is an injective group homomorphism (a group action of $\mathrm{Aut}(\mathbb{K})$ on $\mathbb{K}^n$ by semilinear mappings); in particular, $\{\hat{\sigma};\ \sigma \in \mathrm{Aut}(\mathbb{K})\}$ is a subgroup of $\Gamma L(\mathbb{K}^n)$. If $\hat{\sigma} \in GL(\mathbb{K}^n)$, i.e. $\hat{\sigma}$ linear, then $\hat{\sigma}(\lambda x) = \lambda^\sigma x^\sigma$ but also $\hat{\sigma}(\lambda x) = \lambda\hat{\sigma}(x)$ for all $x \in \mathbb{K}^n$ and $\lambda \in \mathbb{K}$, so is $\lambda^\sigma = \lambda$ and so $\sigma = \mathrm{id}$. Instead of $\hat{\sigma}$ we will now write σ again and consider $\sigma \in \mathrm{Aut}(\mathbb{K})$ also as a semilinear mapping on $\mathbb{K}^n$.

(b) For a given semilinear mapping S we have $S(\lambda x) = \lambda^\sigma Sx$ for some $\sigma \in \mathrm{Aut}(\mathbb{K})$. Let $\tau = \sigma^{-1}$. Then $A := S \circ \tau = S\tau$ is linear, because $A(\lambda x) = S((\lambda x)^\tau) = S(\lambda^\tau x^\tau) = (\lambda^\tau)^\sigma S(x^\tau) = \lambda A(x)$. So we have $S = A\sigma$ for some $A \in GL(\mathbb{K}^n)$ and $\sigma \in \mathrm{Aut}(\mathbb{K})$.

Uniqueness: When $S = A\sigma = B\rho$ with $A, B \in GL(\mathbb{K}^n)$ and $\sigma, \rho \in \mathrm{Aut}(\mathbb{K})$, then $\rho\sigma^{-1} = B^{-1}A \in \mathrm{Aut}(\mathbb{K}) \cap GL(\mathbb{K}^n) = \{\mathrm{id}\}$ according to (a). Thus $\rho = \sigma$ and $A = B$.

(c) For $A, B \in GL(\mathbb{K}^n)$ and $\alpha, \beta \in \text{Aut}(\mathbb{K})$ we have

$$A\alpha B\beta = A\alpha B\alpha^{-1}\alpha\beta = AB^{\alpha}\alpha\beta$$

with $B^{\alpha} := \alpha B\alpha^{-1}$. This is a linear map whose matrix is obtained from applying the automorphism α to all matrix coefficients of B: In fact, the i-th column of B^{α} is $B^{\alpha}e_i = \alpha(B(\alpha^{-1}e_i)) = \alpha(Be_i)$, so α is applied to each component of the i-th column of B. Thus, the product is equal to that in the semidirect product of $GL(\mathbb{K}^n)$ and $\text{Aut}(\mathbb{K})$,

$$(A, \alpha) \cdot (B, \beta) = (AB^{\alpha}, \alpha\beta).$$

18. The initial step for $k = 1$ is clear by definition of a collineation.
Induction step: We wish to prove the assertion for a k-dimensional subspace $P' \subset P$ with $k \geq 2$. We span P' by a $(k-1)$-dimensional subspace $P'' \subset P'$ and a transversal straight line $g \subset P'$, that is $P' = P'' \vee g$, i.e. the points of P' lie on the straight lines connecting points of P'' with g. The intersection of the subspace P'' with the straight line g is a point $s \in P'$. By induction hypothesis, $F(P'')$ is again a $(k-1)$-dimensional subspace. Moreover, since F is a collineation, $F(g)$ is a straight line connecting the subspace $F(P'')$ with the point $F(s)$.
Claim: $F(P') \subset F(P'') \vee F(g)$. We have to show $F(p) \in F(P'') \vee F(g)$ for any $p \in P'$. If $p \in P''$ or $p \in g$, this is clear. Otherwise we connect p with a point $q \in g$ by a straight line h which is contained in $P' = P'' \vee g$ and meets the hyperplane $P'' \subset P'$ at a point r. The image $F(h)$ is the straight line through the points $F(q) \in F(g)$ and $F(r) \in F(P'')$, therefore $F(h) \subset F(P'') \vee F(g)$ and so $F(p) \in F(P'') \vee F(g)$, hence $F(P'' \vee g) \subset F(P'') \vee F(g)$.
Likewise, the collineation F^{-1} satisfies $F^{-1}(F(P'') \vee F(g)) \subset P'' \vee g = P'$, and thus the other inclusion follows: $F(P'') \vee F(g) \subset F(P')$, that is, we have equality. Therefore, the image of P' is the k-dimensional subspace $F(P'') \vee F(g)$.
19. The equations $x^2 \pm y^2 - 1 = 0$ are homogenized to $x^2 \pm y^2 - z^2 = 0$, thus to (1): $x^2 + y^2 - z^2 = 0$ and (2): $x^2 - y^2 - z^2 = 0$. Further, $y - x^2 = 0$ becomes (3): $yz - x^2 = 0$. By swapping the x- and the z-coordinate, i.e. by the linear substitution of variables $x = \tilde{z}$, $y = \tilde{y}$, $z = \tilde{x}$, the equation (1) becomes $\tilde{z}^2 + \tilde{y}^2 - \tilde{x}^2 = 0$, hence $\tilde{x}^2 - \tilde{y}^2 - \tilde{z}^2 = 0$ which is $(\tilde{2})$. To (3), $yz - x^2 = 0$, we apply the linear substitution $y = \tilde{z} + \tilde{y}$, $z = \tilde{z} - \tilde{y}$, $x = \tilde{x}$ and obtain $\tilde{z}^2 - \tilde{y}^2 - \tilde{x}^2 = 0$ or $\tilde{x}^2 + \tilde{y}^2 - \tilde{z}^2 = 0$, which is $(\tilde{1})$.
20. From $F[e_i] = [a_i]$ and $F[e] = [a]$ we obtain $Ae_i = \lambda_i a_i$ and $Ae = \lambda a$ with $\lambda_i, \lambda \in \mathbb{K}^*$. As A can be modified by a nonzero scalar factor without changing $F = [A]$, we may choose $\lambda = 1$. From $Ae = a$ the λ_i can be calculated: On the one hand, $a_1, \ldots, a_{n+1}$ is a basis and therefore $a = \sum_i \mu_i a_i$ where the $\mu_i \in \mathbb{K}^*$ are considered to be known, since a and a_i are known. (By assumption, no μ_i

can be zero.) On the other hand $Ae = \sum_i Ae_i = \sum_i \lambda_i a_i$. From $Ae = a$ we obtain $\sum_i \lambda_i a_i = \sum_i \mu_i a_i$ and thus $\lambda_i = \mu_i$ since the vectors $a_1, \ldots, a_{n+1}$ are linearly independent. Therefore A is determined.

21. The homogenization of the hyperbolic equation $x^2 - y^2 = 1$ is $x^2 - y^2 - z^2 = 0$. The affine part $\{z = 1\}$ is the affine hyperbola $x^2 - y^2 = 1$. The projective curve $\{[x, y, z];\ x^2 - y^2 - z^2 = 0\}$ also intersects the line at infinity $\{z = 0\}$, namely at the two points $x = y, z = 0$ and $x = -y, z = 0$. These are the points at infinity of the (projective closures of the) two asymptotes $x = y, z = 1$ and $x = -y, z = 1$. The equation $x^2 - y^2 - z^2 = 0$ describes a cone in space whose axis is the x-axis. It consists of a bundle of straight lines through O, the generatrices, all but two of which intersect the affine plane $\{z = 1\}$, namely at the points of the affine hyperbola $x^2 - y^2 = 1$, $z = 1$. The exceptions are the two generatrices which run entirely in the xy-plane $\{z = 0\}$: the straight lines $x = y, z = 0$ and $x = -y, z = 0$.

   You can also see this in the plane itself without passing to space: The connecting straight lines from any chosen point (e.g., the origin) to the points P on a hyperbola branch tend towards a (parallel of an) asymptote as P tends to infinity. This shows that the point at infinity associated to the class of lines parallel to the asymptote must lie in the projective closure of the hyperbola, see Footnote 8 in Sect. 3.2.

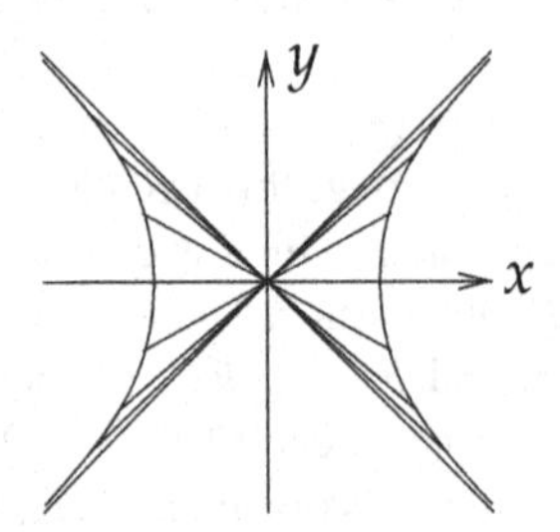

22. The quadric is projectively equivalent to the one-sheeted hyperboloid, because the normal form of the defining quadric contains two minus signs and two plus signs. We see this, for example, by simultaneous row and column transformations:

$$\begin{array}{rrrr|rrrr|rrrr|rrrr}
1 & 2 & 0 & 3 & 1 & 0 & 0 & 0 & 1 & 0 & 0 & 0 & 1 & 0 & 0 & 0 \\
2 & 2 & 0 & 4 & 0 & -2 & 0 & -2 & 0 & -2 & 0 & 0 & 0 & -2 & 0 & 0 \\
0 & 0 & 1 & -1 & 0 & 0 & 1 & -1 & 0 & 0 & 1 & -1 & 0 & 0 & 1 & 0 \\
3 & 4 & -1 & 4 & 0 & -2 & -1 & -5 & 0 & 0 & -1 & -3 & 0 & 0 & 0 & -4
\end{array}$$

23. The quadric is $Q = \{[x];\ x \in \mathbb{R}^4,\ q(x) = 0\}$ with $q(x) = q(s, t, u, v) = st - uv$. The tangent plane is $T_{[x]}Q = \{[y];\ \beta(x, y) = 0\}$ where β is the

bilinear form belonging to the quadratic form q using polarization:[1] $2\beta(x, y) = q(x + y) - q(x) - q(y)$. Thus a straight line $g \subset Q$ through a point $[x] \in Q$ lies also in $T_{[x]}Q$: For all $[y] \in g$ we have $[x + y] \in g$ (since $g = \pi(E)$ for a two-dimensional linear subspace E containing x, y and hence $x + y$). So are $[x], [y], [x + y] \in g \subset Q$, and so is $q(x) = q(y) = q(x + y) = 0$ and $\beta(x, y) = 0$, thus $[y] \in T_{[x]}Q$.

24. The previously parallel straight lines in the figure of Exercise 4 must now meet at points that lie on a common straight line.
25. *Pappus:* Given six points $A, \ldots, F$ lying alternately on two straight lines a and b. Let XY denote the connecting line of two points X, Y. Then the points $AB \wedge DE$, $BC \wedge EF$, $CD \wedge FA$ lie on a common straight line c.
    Dual: Given six straight lines $a, \ldots, f$ passing alternately through two points A and B. Let xy denote the intersection of two straight lines x, y. Then the lines $ab \vee de$, $bc \vee ef$, $cd \vee fa$ pass through a common point C.

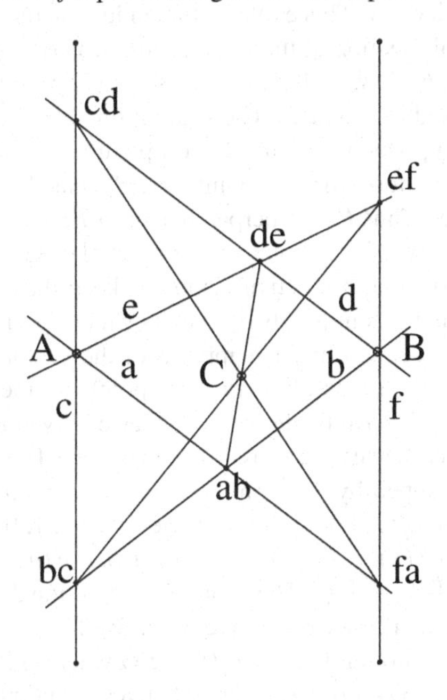

Pappus' theorem is a special case of Pascal's theorem when a degenerate conic section (a pair of straight lines) is also admitted there. The theorem dual to Pascal's is Brianchon's, but in dualizing we use that the conic section is non-degenerate: The tangents of the conic section with symmetric bilinear form β are dual to the points of the conic section with bilinear form β^{-1}, see Lemma 3.7. Therefore the dual Pappus theorem cannot be a special case of Brianchon's. However it is a limiting case: The conic section in Brianchon's

[1] Here we need to assume $\operatorname{char}(\mathbb{K}) \neq 2$.

theorem can be a very narrow ellipse such that the points of contact of the tangents are alternately very close to the two endpoints A and B of the long axis, like in the right figure before Theorem 3.6. Then we obtain the dual Pappus theorem as a limiting case where the ellipse degenerates to the line segment $[A, B]$ and the points of contact of the tangents become closer and closer to A and B. However, the line segment $[A, B]$ itself is not the solution set of a quadratic equation, unlike the pair of straight lines in Pappus' theorem.

26. The polar to a point $[x]$ consists of the homogeneous vectors $[y]$ which are perpendicular to x with respect to the given symmetric bilinear form β, that is $\beta(x, y) = 0$. If $Q = \{[x];\ \beta(x, x) = 0\}$ denotes the corresponding conic section, then the tangent at the point $P = [x] \in Q$ is the straight line

$$T_{[x]}Q = \{[y];\ \beta(x, y) = 0\}$$

which is the polar to x. This explains the middle picture. If a point $P = [y]$ lies outside the conic section at the intersection of two tangents $T_{[x]}Q$ and $T_{[x']}Q$ then $\beta(x, y) = \beta(x', y) = 0$ and thus also $\beta(x'', y) = 0$ for all x'' in the linear subspace spanned by x and x'. The straight line $g = [x] \vee [x']$ is therefore the polar to $P = [y]$ which explains the left picture. If P lies in the interior of the conic section, we look for two points P_1, P_2 outside whose polars g_1, g_2 are intersecting at P. Thus P is β-perpendicular to P_1 and P_2 and thus to all points of $g = P_1 \vee P_2$, so g is the polar to P. This explains the right picture.

27. We already know that projective mappings keep the cross-ratio invariant: We always represent the four points by vectors which lie on a common straight line in the vector space, and linear mappings of the vector space preserve straight lines as well as the ratios of every three points on the straight line, thus also their quotients. We have to show the converse: Given a bijective mapping $F : \mathbb{P}^1 \to \mathbb{P}^1$ which leaves the cross-ratio invariant. The three points $0, 1, \infty \in \hat{\mathbb{K}} = \mathbb{P}^1$ are mapped by F^{-1} onto any three points $a, b, c \in \hat{\mathbb{K}}$: $F(a) = 0$, $F(b) = 1$, $F(c) = \infty$. For each $y \in \mathbb{K}$ we have $(y, 1; 0, \infty) = \frac{y-0}{1-0} \cdot \frac{1-\infty}{y-\infty} = y$. Thus $F(x) = (F(x), 1; 0, \infty) = (F(x), F(b); F(a), F(c)) = (x, b; a, c) = \frac{x-a}{b-a} \cdot \frac{b-c}{x-c} = \frac{x(b-c)-a(b-c)}{(b-a)x-(b-a)c}$. Therefore F is a fractional-linear function (Möbius transformation) and hence projective (Exercise 15).

An alternative proof is to pass to $H = FG$ where G is the projective mapping which maps $0, 1, \infty$ onto a, b, c. Then H fixes the points $0, 1, \infty$ and preserves the cross-ratio, and since $(x, 1; 0, \infty) = x$, we have $H(x) = x$ for all x, thus $F = G^{-1}$.

28. (a): For reasons of symmetry, the diagonals are parallel to the sides: Since sides and diagonals are preserved under the reflections of the pentagon, they are both perpendicular to an axis of symmetry, because their endpoints are interchanged by the reflection. So the sides of the two hatched triangles are parallel, therefore the angles are equal.

29. The quadrilateral *abde* is a parallelogram, even a rhombus: All four sides are equal in length, namely equal to the radius, which is the same for all circles, and which we set equal to one. Since the line *ab* is horizontal, the points e, d, f

are on a common horizontal line perpendicular to the vertical line dc. Thus edg is an isosceles right triangle and the straight line $eg = ei$ has slope 45°. The x- and y-coordinates of the difference vector $i - e$ should therefore be equal.

On the other hand, we can compute these components assuming that the construction yields a regular pentagon with side length 1 and diagonal $|i - h|$. Then $|i - h| = \Phi$, and $(i - e)_x = \frac{1}{2}|i - h| + |d - e| = \frac{\Phi}{2} + 1 \approx 1{,}809$. For the y-component, we must first calculate the height h of the horizontal diagonal above the base of the pentagon.

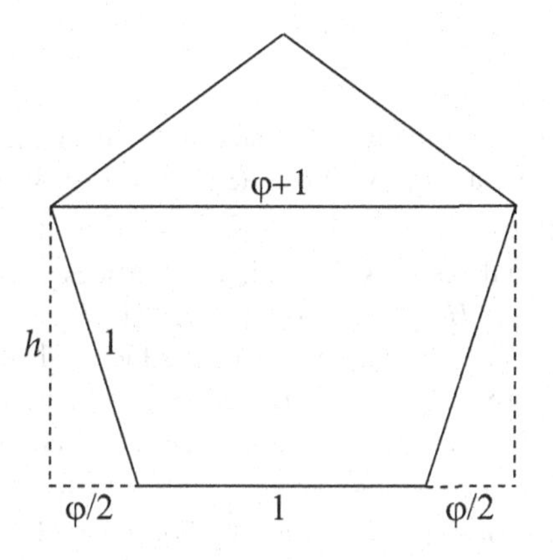

Thus $h^2 = \frac{1}{4}(4 - \varphi^2) = \frac{1}{4}(3 + \varphi)$ (using $\varphi^2 = 1 - \varphi$). To h we must add the height of the equilateral triangle abd with side length 1 which is $\sqrt{3}/2$. Thus $(i - e)_y = (i - b)_y + (a - d)_y = \frac{1}{2}(\sqrt{3 + \varphi} + \sqrt{3}) \approx 1{,}817 \neq 1{,}809 \approx (i - e)_x$. But even without further calculation you can see that the two numbers are different, because the number $(i - e)_y$ contains $\sqrt{3}$, but not so $(i - e)_x$, and the square roots $\sqrt{3}$ and $\sqrt{5}$ (occurring in Φ) are linearly independent over the rational numbers. So Dürer's construction is just a (pretty good) approximation to a pentagonal construction.

30. According to Pythagoras we have $|A - B|^2 = \frac{1}{4} + 1 = \frac{5}{4}$, thus $|C - M| = \frac{1}{2}(\sqrt{5} - 1) = \varphi$, and hence

$$s = |C - A| = \sqrt{1 + \varphi^2} = \sqrt{2 - \varphi}.$$

We must show that this is the side length of the pentagon inscribed in the unit circle (right figure in the exercise). The two colored right triangles are similar because they both have an angle of 36°. This can be seen from the left figure below, which shows that the three angles, that add up to the interior angle 108° of the pentagon, are equal, i.e. 36°. It follows also from the inscribed angle theorem (see end of Sect. 3.8) which states $2(\alpha + \beta) = \gamma$ in the right figure below; in our case $\gamma = 360°/5 = 72°$, hence $\alpha + \beta = 36°$.

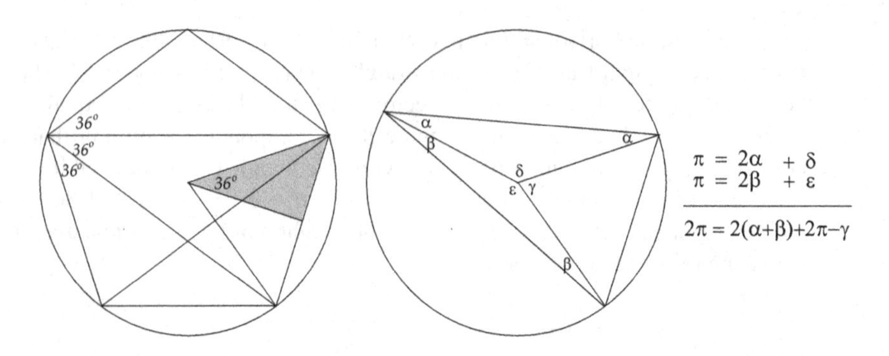

The rest of the argument is as stated in the exercise: From the right colored triangle, using Pythagoras, we take : $(s/2)^2 = 1-(\Phi/2)^2 = \frac{1}{4}(4-(\varphi+1)^2) = \frac{1}{4}(3-\varphi^2-2\varphi) = \frac{1}{4}(3-(1-\varphi)-2\varphi) = \frac{1}{4}(2-\varphi)$.

31. The n-dimensional simplex Σ with the vertices $e_1, \dots, e_{n+1} \in \mathbb{R}^{n+1}$ lies in the hyperplane $H = \{x \in \mathbb{R}^{n+1};\ \langle x, d\rangle = 1\} = e_1 + d^\perp$ with $d = (1, \dots, 1)$. The normal vectors of the faces whose angles we are looking for must therefore lie in $d^\perp$. The vector e_i is perpendicular to the face Σ_i with the vertices $e_1, \dots, e_{i-1}, e_{i+1}, \dots, e_{n+1}$, ; its component in $d^\perp$ is $v_i = e_i - \frac{\langle e_i, d\rangle}{\langle d, d\rangle} d = e_i - \frac{1}{n+1} d = \frac{1}{n+1}(-1, \dots, n, \dots, -1)$ and consequently $|v_i|^2 = (n^2+n)/(n+1)^2 = n/(n+1)$. For the angle β_n between two normals, e.g. v_1 and v_2, the result is $\cos\beta_n = \frac{\langle v_1, v_2\rangle}{|v_1||v_2|}$. Thus for the *dihedral angle* between the hyperplanes, $\alpha_n = 180° - \beta_n$, we obtain $\cos\alpha_n = -\cos\beta_n = -\frac{1}{n(n+1)}\langle(n, -1, -1, \dots), (-1, n, -1, \dots)\rangle = -\frac{-2n+n-1}{n(n+1)} = \frac{1}{n}$. Hence $\alpha_2 = 60°$, $\alpha_3 \approx 70.5°$, $\alpha_4 \approx 75.5°$ and $\alpha_n \nearrow 90°$ for $n \to \infty$.

Consequence: On the boundary of a 4-dimensional Platonic solid, three, four, or five 3-simplices (tetrahedra) are adjacent to a one-dimensional face (edge), because $5 \cdot 70.5° < 360° < 6 \cdot 70.5°$. But in dimension $n = 5$, only three or four 4-simplices can be adjacent to a 2-dimensional face, because $5 \cdot 75.5° > 360°$. Thus there are only two Platonic 5-solids bounded by 4-simplices: 5-simplex and co-cube. The dihedral angles of the other four-dimensional Platonic solids are 90° for the cube, 120° for the cocube,[2] and > 120° for the remaining ones. Thus five-dimensional Platonic solids are bounded either by 4-simplices (three or four around a common 2-face, bounding 5-simplex and 5-cocube) or by 4-cubes (three around a common 2-face, bounding the 5-cube). Since the dihedral angles increase with dimension, the analogous facts holds for any $n \geq 5$.

[2] The dihedral angle of the cocube was calculated at the end of Sect. 4.5 for $n = 3$. Analogously we obtain $\cos\beta = (n-2)/n$ for arbitrary dimension n. For $n = 4$ we have $\cos\beta = 1/2$ and thus $\beta = 60°$ and $\alpha = 120°$. Hence three four-dimensional co-cubes around a common 2-face leave no gap and tessellate $\mathbb{R}^4$; this is related to the lattice F_4, cf. https://en.wikipedia.org/wiki/F4_(mathematics).

32. For reasons of symmetry (?!) the following calculations are sufficient:
   (a) $|(\varphi, \Phi, 0)|^2 = \varphi^2 + \Phi^2 = 1 - \varphi + 1 + \varphi + 1 = 3 = |(1, 1, 1)|^2$.
   (b) $|(\varphi, \Phi, 0) - (-\varphi, \Phi, 0)| = 2\varphi$,
   $|(\varphi, \Phi, 0) - (1, 1, 1)|^2 = |(\varphi - 1, \Phi - 1, -1)|^2 = |(\varphi - 1, \varphi, -1)|^2 = \varphi^2 - 2\varphi + 1 + \varphi^2 + 1 = 2(\varphi^2 - \varphi + 1) = 4\varphi^2$.
   (c) $(\Phi, 0) - (\varphi, \Phi) = (1, -\Phi)$,
   $(\Phi, 0) - (1, 1) = (\Phi - 1, -1) = (\varphi, -1) = \varphi(1, -\Phi)$.
   (d) diagonal/side $= |(1, 1, 1) - (-1, 1, 1)|/(2\varphi) = 2/(2\varphi) = \Phi$.
33. (a) It is $|2e_i| = 2$ and $|(\pm 1, \dots, \pm 1)| = \sqrt{n}$. These norms are equal precisely for $n = 4$. Only in this dimension, the vertices of the cube and the cocube are all lying on a common sphere.
   (b) The edges of K are the edges of the cube together with the "connecting edges", the connections between cocube vertices and adjacent cube vertices. Adjacent to the cocube vertex $2e_1$ are the cube vertices $(1, \pm 1, \pm 1, \pm 1)$ (all others have larger distance from $2e_1$), and the distance is $|(2, 0, 0, 0) - (1, \pm 1, \pm 1, \pm 1)| = |(1, \pm 1, \pm 1, \pm 1)| = \sqrt{4} = 2$. Since the edge of the cube has the same length $|(1, 1, 1, 1) - (-1, 1, 1, 1)| = 2$, all edges have equal length.
   (c) Indeed, the reflection along the perpendicular bisector of a connecting edge leaves the set of these 16 + 8 vertices invariant. It suffices (why?) to consider the edge between $p = (2, 0, 0, 0)$ and $q = (1, -1, -1, -1)$. Since the lengths p and q are equal, the perpendicular bisector of the line segment $[p, q]$ passes through the origin. Therefore it is the linear hyperplane $(p - q)^\perp = d^\perp$ with $d = (1, 1, 1, 1)$. The reflection along this hyperplane is $S_d(x) = x - 2\frac{\langle x, d\rangle}{\langle d, d\rangle} d = x - \frac{1}{2}(\sum_i x_i)d$. Since this linear map commutes with all permutations of the four coordinates and, of course, with $-\mathrm{id}$ and these mappings leave the set of vertices invariant, we only need to look at the vertices up to a common sign and arbitrary permutations of the coordinates. It is therefore sufficient to apply S_d to the four vertices $p = 2e_1$, $d = (1, 1, 1, 1)$, $q' = (1, 1, 1, -1)$, and $q'' = (1, 1, -1, -1)$. We get $S_d(p) = p - d = (1, -1, -1, -1) = q$ (which is clear since we reflect along the perpendicular bisector of $[p, q]$), further $S_d(d) = d - 2d = -d$ (also clear), $S_d(q') = q' - d = -2e_4$ and $S_d(q'') = q''$ (of course, since $q'' \perp d$). So the vertex set is invariant under S_d.
   (d) Thus the isometry group G of K acts transitively on the set of vertices, the 16 + 8 vertices of the cube and the cocube. In fact, the common isometry group of cube and cocube is generated by the coordinate permutations and the sign changes of each coordinate and acts transitively on the two vertex sets. Further, S_d maps the cocube vertex $2e_1$ onto the cube vertex $(1, -1, -1, -1)$ and thereby connects the two vertex sets to a single orbit of G. In fact, we can map all 24 octahedra faces onto each other while still rotating each octahedron in all possible 24 ways (size of the rotation group of the octahedron or cube); hence, the rotation group of the 24-cell has order $|G| = 24 \cdot 24$.

34. (b) Let $A = \begin{pmatrix} a & c \\ b & d \end{pmatrix} \in SU(2)$. The two columns are perpendicular to each other. Since $\begin{pmatrix} a \\ b \end{pmatrix}^{\perp}$ is one-dimensional and $\left\langle \begin{pmatrix} -\bar{b} \\ \bar{a} \end{pmatrix}, \begin{pmatrix} a \\ b \end{pmatrix} \right\rangle = -ba + ab = 0$, we have $\begin{pmatrix} c \\ d \end{pmatrix} = \lambda \begin{pmatrix} -\bar{b} \\ \bar{a} \end{pmatrix}$ for some $\lambda \in \mathbb{C}$. Then $1 = \det \begin{pmatrix} a & c \\ b & d \end{pmatrix} = \lambda(a\bar{a} + b\bar{b}) = \lambda$ (because the first column $\begin{pmatrix} a \\ b \end{pmatrix}$ has length one), so $\lambda = 1$.

The inverse also follows: We have $A = \begin{pmatrix} a & -\bar{b} \\ b & \bar{a} \end{pmatrix}$ with $|a|^2 + |b|^2 = 1$ if and only if the columns of A form a unitary basis with determinant one. The subset $\mathbb{H} \subset \mathbb{C}^{2\times 2}$ obviously forms a real linear subspace, even a subalgebra, since we have $\mathbb{H} = \mathbb{R} \cdot SU(2)$, and for any $t_1, t_2 \in \mathbb{R}$ and $A_1, A_2 \in SU(2)$, the product $t_1 t_2 A_1 A_2$ is again in $\mathbb{R} \cdot SU(2)$. This algebra is non-commutative because $SU(2)$ is non-commutative; e.g.,

$$\begin{pmatrix} & i \\ i & \end{pmatrix} \begin{pmatrix} & -1 \\ 1 & \end{pmatrix} = \begin{pmatrix} i & \\ & -i \end{pmatrix} = -\begin{pmatrix} & -1 \\ 1 & \end{pmatrix} \begin{pmatrix} & i \\ i & \end{pmatrix}.$$

Since every element tA with $t \neq 0$ has the inverse $t^{-1}A^{-1}$, the algebra $\mathbb{H} = \mathbb{R} \cdot SU(2)$ is a skew field.

35. (a) The focal points of the ellipse are two points F, F' on the long axis with the property that $|F - P| + |F' - P| = c = const$ for each point P of the ellipse. Let us set for P a point of the ellipse on the long axis, then we see $|F - P| + |F' - P| = 2a$, that is $c = 2a$. On the other hand, let us set for P an ellipse point on the short axis, then we get $|F - P| = |F' - P| = c/2 = a$. The triangle (F, O, P) is then right-angled, so Pythagoras says $f^2 = |F|^2 = |F - P|^2 - |P|^2 = a^2 - b^2$.

(b) The focal points F, F' of the hyperbola lie on the axis intersected by the hyperbola, and they have the property that $|F - P| - |F' - P| = c = const$ for each point P of the hyperbola. Choosing for P the intersection of the right hyperbolic branch with the axis, we obtain $c = 2a$.

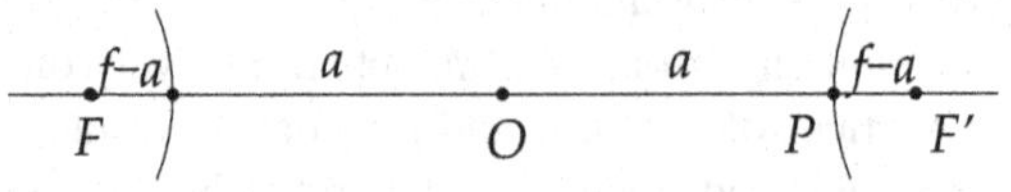

f–a a a f–a F O P F′

On the other hand, if we let the point P on this branch of the hyperbola go to infinity, the straight lines FP and $F'P$ are more and more parallel to one of the asymptotes (see the figure below, a detail from the right-hand figure in the exercise). Since the difference in length always remains $c = 2a$, the parallels to an asymptote that pass through the two focal points also have length difference c, see subsequent figure. Thus, the parallel through F together with its perpendicular through F' and the x-axis form a right triangle with a cathetus of length $2a$ and hypotenuse with length $|F' - F| =$

$2f$. The second cathetus must then have length $2b$, because the triangle is similar (same angles) to the one with vertex 0 and catheti a and b. Therefore, according to Pythagoras, $f^2 = a^2 + b^2$.

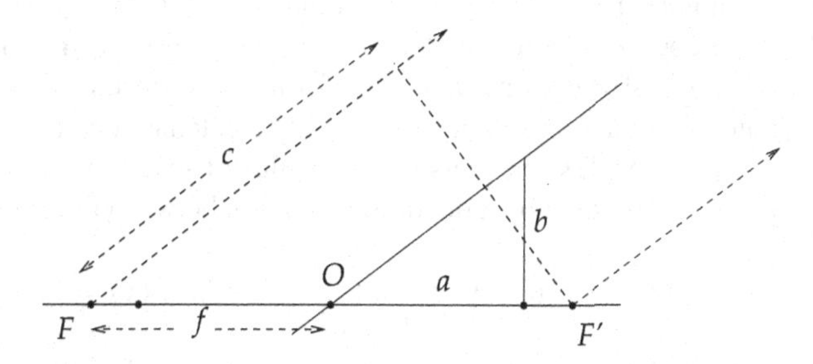

36. Let h be the height of P above (or better below) B, the plane of the circle of contact. Then the shortest line segment from P to the directrix $g = E \cap B$ includes the angle β with the horizontal, so its length is equal to $h/\sin\beta$. The generatrix m of the cone through the point P includes the angle α with the horizontal, so the segment between P and B on the generatrix equals $h/\sin\alpha$. Hence the ratio of these two distances is

$$(h/\sin\alpha)/(h/\sin\beta) = \sin\beta/\sin\alpha = const.$$

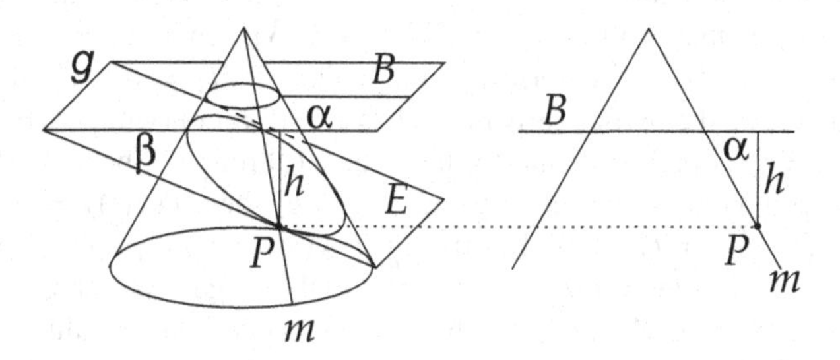

37. The angle 7.2° is 1/50 of the full angle 360°. The angle 7.2° corresponds to the distance $d = 800$ km, hence the full angle $360° = 50 \cdot 7.2°$ corresponds to $50 \cdot 800 = 40\,000$ km; this is the circumference of the Earth.

    How much do the 3° of deviation from the meridian account for? Neglecting the fact that the circle of latitude for Aswan at 24° is shorter than the equator by a factor $\cos 24° = 0{,}91$ (which would make the error even smaller), we obtain that the diagonal in the rectangle with sides 7.2 and 3 according to Pythagoras is equal to 7.8, hence it is a good 8 % longer than the side. But the measurements of such large distances were probably subject to even larger errors at those days.
38. With $w := \sqrt{1+(f')^2}$ we obtain $g_{st} = h_{st} = 0$, $g_{ss} = w^2$, $g_{tt} = f^2$, $h_{ss} = -f''/w$, $h_{tt} = f/w$ (please recalculate!) and from this $\kappa_1 = -f''/w^3$ and $\kappa_2 = 1/(fw)$. The minimum area equation $H = 0$ or $-\kappa_1 = \kappa_2$ then results in $f'' = w^2/f = (1+(f')^2)/f$. This is a *differential equation*, an

equation between a sought function f and its derivatives f', f''. A solution of this equation is $f = \cosh$ with $f' = \sinh$ and $f'' = \cosh$, because $(1 + (f')^2)/f = (1+\sinh^2)/\cosh = \cosh^2/\cosh = \cosh = f''$. All other solutions arise from this by homotheties of the graph $\{(x, y) \in \mathbb{R}^2;\ y = \cosh x\}$. The graph of the cosh-function is also called *catenary*, since a free-hanging chain with links of equal weight assumes this form; the surface of revolution of this curve is therefore called a *catenoid* (from Latin catena = chain).

39. First we have to recall some facts from calculus of several variables. A quadric is a hypersurface defined by a quadratic equation between the coordinates:

$$Q = f^{-1}(c) = \{f = c\} = \{x \in \mathbb{R}^n;\ f(x) = c\},$$

where c is a real constant and f is a quadratic (order 2) polynomial. In other words, Q is the set of solutions x of the equation $f(x) = c$, also called *level set* for the level c. But such set is not always a hypersurface. E.g., for $n = 2$ and $f(x) = x_1 x_2$, the level set $\{f = 0\}$ is the coordinate cross, the union of x_1- and x_2- axes, because $x_1 x_2 = 0 \iff x_1 = 0$ or $x_2 = 0$. This set is not a hypersurface in $\mathbb{R}^2$ (a regular curve), because at the origin it cannot be approximated by a hyperplane (a straight line). However, there is a sufficient criterion when a level set $L = \{f = c\}$ *is* a hypersurface: This holds if the derivative or the *gradient* of f is nonzero at any point $x \in L$ (*inverse function theorem*). For a function $f : \mathbb{R}^n_o \to \mathbb{R}$ the derivative df_x at any point x is a *linear form*, that is a linear map of $\mathbb{R}^n$ to $\mathbb{R}$. Written as a matrix it is a row, and the *gradient* ∇f_x is the corresponding column: $\nabla f_x = (df_x)^T$. Its components are therefore the partial derivatives $f_1(x), \dots, f_n(x)$ (with $f_i := \partial f/\partial x_i$). This vector ∇f_x is perpendicular to the level set through x, because if a curve $x(t)$ is entirely contained in a level set $\{f = c\}$ then $f(x(t)) = c$ for all t and therefore $\frac{d}{dt} f(x(t)) = 0$, thus $0 = \frac{d}{dt} f(x(t)) = df_{x(t)} x'(t) = \langle \nabla f_{x(t)}, x'(t)\rangle$. The gradient is therefore perpendicular to the tangent vector $x'(t)$ of any curve in the level set L. So much for the general theory from calculus as background.

In our case the defining function f is the quadratic form $q^u(x) = \sum_i \frac{x_i^2}{a_i - u}$ and the level hypersurfaces are the quadrics $Q_u = \{q^u = 1\}$. The i-th partial derivative is $(q^u)_i(x) = \frac{2x_i}{a_i - u}$ and thus

$$\nabla(q^u)_x = \left(\frac{2x_1}{a_1 - u}, \dots, \frac{2x_n}{a_n - u}\right)^T \neq 0 \text{ for all } x \neq 0,$$

and in particular this holds for all $x \in Q_u$. Furthermore

$$\langle \nabla(q^u)_x, \nabla(q^v)_x \rangle = 4 \sum_i \frac{x_i^2}{(a_i - u)(a_i - v)},$$

$$\begin{aligned} q^u(x) - q^v(x) &= \sum_i \left(\frac{x_i^2}{a_i - u} - \frac{x_i^2}{a_i - v}\right) \\ &= \sum_i \frac{x_i^2(u - v)}{(a_i - u)(a_i - v)} \\ &= (u - v) \sum_i \frac{x_i^2}{(a_i - u)(a_i - v)}. \end{aligned}$$

From this results, the claimed equation $4(q^u - q^v) = (u - v)\langle \nabla q^u, \nabla q^v \rangle$ follows for all $u, v \in \mathbb{R} \setminus \{a_1, \ldots, a_n\}$. So if $x \in Q_u \cap Q_v$ for $u \neq v$, then $q^u(x) = q^v(x) = 1$ and thus $\nabla(q^u)_x \perp \nabla(q^v)_x$. So the hypersurfaces Q_u and Q_v intersect orthogonally at x, because ∇q^u and ∇q^v are normal vector fields along Q_u and Q_v.

It only remains to consider for which pairs of parameters u, v the corresponding quadrics intersect. For this purpose we discuss the function $u \mapsto q^u(x)$ for fixed $x \in \mathbb{R}^n$ with $x_i \neq 0$ for all $i = 1, \ldots, n$. It has poles at $a_1, \ldots, a_n$ and its derivative is positive everywhere else: $\frac{\partial}{\partial u} q^u(x) = \sum_i \frac{x_i^2}{(a_i - u)^2} > 0$. For $u \to \pm\infty$ we have $q^u(x) \to 0$. In the interval $I_1 = (-\infty, a_1)$ the function increases from 0 to ∞ strictly monotonically, and in each of the intervals $I_i = (a_{i-1}, a_i)$ for $i = 2, \ldots, n$ it increases strictly monotonically from $-\infty$ to ∞. In the interval $I_{n+1} = (a_n, \infty)$ it increases from $-\infty$ to 0, so it is negative there. Conclusion: In each of the intervals $I_1, \ldots, I_n$ the value 1 is assumed exactly once: To every such x there is exactly one $u_i \in I_i$ with $q^{u_i}(x) = 1$, that is $x \in Q_{u_i}$. In other words: Through each point $x \in \mathbb{R}^n_o = \{x;\ x_i \neq 0\ \forall i\}$ there is exactly one quadric from each of the n families $(Q_u)_{u \in I_i}$, $i = 1, \ldots, n$, and there the normal vectors are perpendicular to each other.

In the case $n = 2$ we have $I_1 = (-\infty, a_1)$ and $I_2 = (a_1, a_2)$. For each $u \in I_1$ the numbers $a_1 - u$ and $a_2 - u$ are both positive, and the conic section $Q_u = \{(x, y) \in \mathbb{R}^2;\ \frac{x^2}{a_1 - u} + \frac{y^2}{a_2 - u} = 1\}$ is an ellipse with short semi-axis $b = \sqrt{a_1 - u}$ and long semi-axis $a = \sqrt{a_2 - u}$. According to Exercise 35 all these ellipses have the same focal points, because $f^2 = a^2 - b^2 = a_2 - a_1$ is independent of u. Thus, the ellipses all have the same foci; they are *confocal*. If $u \in I_2$ then $a_1 - u < 0 < a_2 - u$, hence $Q_u = \{-\frac{x^2}{u - a_1} + \frac{y^2}{a_2 - u} = 1\}$ is a hyperbola opening upwards and downwards with semi-axes $a = \sqrt{u - a_1}$ and $b = \sqrt{a_2 - u}$. The distance f of the focal points from the center is (according to Exercise 35) independent of u, because $f^2 = a^2 + b^2 = u - a_1 + a_2 - u = a_2 - a_1$. Thus the hyperbolas all have the same focal points, and indeed the same as the ellipses. Ellipses and hyperbolas also intersect perpendicularly, as we have shown for any n (left figure).

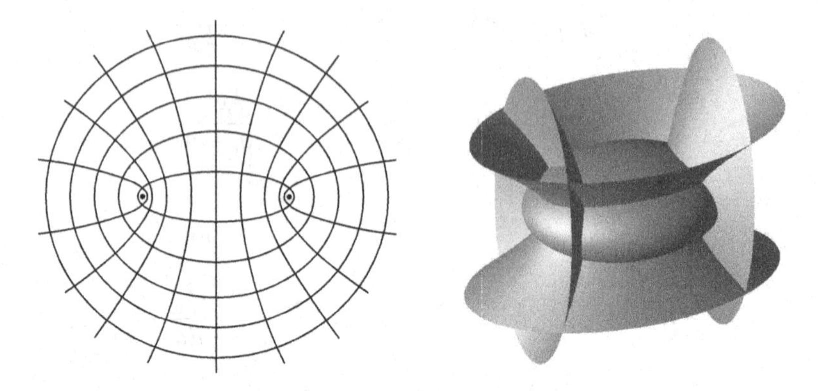

In the case $n = 3$ one has three orthogonal surface families: Ellipsoids and one- and two-sheeted hyperboloids (right figure).[3]

40. (a) The scaling factor (conformal factor) $|d\mu_x(v)|/|v|$ of the conformal mapping μ depends only on the point x and not on the direction v. So it can be calculated for $v = \partial x(\varphi, \theta)/\partial\varphi$. Since the circle of latitude θ is shortened by a factor $\cos\theta$ (compared to the equator), we have $|v| = \cos\theta$. On the other hand, $|d\mu_x v| = 1$ because $\varphi \mapsto \mu(x(\varphi, \theta))$ traverses the line segment $[-\pi, \pi] + i\theta$ with velocity one, according to (1). Hence $\lambda = 1/\cos\theta$.

    (b) According to (1), (2), (3) we have $\mu(x(\varphi, \theta)) = \varphi + if(\theta)$ for some strictly monotonic increasing function $f : (-\pi/2, \pi/2) \to \mathbb{R}$. Thus $\partial\mu/\partial\theta = if'(\theta)$ and on the other hand $|\partial\mu/\partial\theta| = |d\mu.x_\theta| = 1/\cos\theta$ according to (a), since $|x_\theta| = |\partial x/\partial\theta| = 1$. Thus $f' = 1/\cos\theta$ and consequently $f(\theta) = \int_0^\theta (1/\cos t)dt$.

    (c) The map $\alpha(z) = \Phi(e^{i\bar{z}})$ is a composition of conformal mappings. The inner mapping $z = s + it \mapsto e^{i\bar{z}} = e^t e^{is}$ transforms the xy-coordinate system into a polar coordinate system: The horizontal coordinate line $s \mapsto s + it$ is wound around the circle of radius e^t centered at 0, the vertical one $t \mapsto s + it$ is mapped onto the ray $t \mapsto e^{is}e^t$ where the distance to 0 increases monotonically with t and tends to zero for $t \to -\infty$. Under the outer map Φ, the circles centered at 0 become latitudinal circles on the sphere (in particular the circle of radius one is mapped to the equator), and $\partial\varphi(\Phi(e^t e^{is}))/\partial s = 1$. Further, the radial rays become meridians, which for $t \to \pm\infty$ tend to the north and south poles, respectively, and $\partial\theta(\Phi(e^t e^{is}))/\partial t > 0$. Therefore, the inverse image of α has the properties (1), (2), (3), and $\alpha^{-1} = \mu$ because of the uniqueness in (b).

[3] Figure from Eschenburg-Jost [21], Chap. 5.

41. (a) Let $\Phi_\pm : \mathbb{R}^n \to \mathbb{S}^n$ denote the stereographic projection from $\pm N$ (north and south poles). Then $\Phi_- = S \circ \Phi_+$ where $S(x,t) := (x,-t)$ for all $(x,t) \in \mathbb{R}^n \times \mathbb{R}$. Denoting $\Psi_\pm = \Phi_\pm^{-1}$ the inverse maps, we have $\Psi_- = \Psi_+ \circ S$. From Eq. (6.8) follows $\Phi_+(x) = \frac{1}{|x|^2+1}(2x, |x|^2-1)$ and $\Psi_-(w,t) = \Psi_+(w,-t) = w/(1+t)$. Putting $w = 2x/(|x|^2+1)$ and $t = (|x|^2-1)/(|x|^2+1)$, we obtain $1+t = 2|x|^2/(|x|^2+1)$ and hence

$$\Psi_-(\Phi_+(x)) = \frac{w}{1+t} = \frac{x}{|x|^2}.$$

(b) Consider the figure in Exercise 41. The shaded right triangles SOx' and NPS as well as NPS and NOx each have an angle in common, so they are similar. Consequently, SOx' and NOx are similar, so it follows $|x'|/1 = 1/|x|$ and thus $|x'| = 1/|x|$. Moreover, x' and x are pointing in the same direction. Thus the map $x \mapsto x'$ is the inversion along the unit sphere.

42. (a) The straight line g through P is parametrized by $g(t) = P + tv$ for some unit vector v. It meets the circle K at the point $g(t)$ if and only if $|g(t)|^2 = r^2$, which means $|P|^2 + 2t\langle P, v\rangle + t^2 = r^2$ (using $|v|^2 = 1$), in other words, if t is a solution of the quadratic equation

$$t^2 + 2t\langle P, v\rangle t + |P|^2 - r^2 = 0.$$

The solutions t_1, t_2 of any quadratic equation $t^2 + at + b = 0$ satisfy $t_1 + t_2 = -a$ and $t_1 t_2 = b$, which can be seen by comparing coefficients: $t^2 + at + b = (t - t_1)(t - t_2) = t^2 - (t_1 + t_2)t + t_1 t_2$.[4] If $A_1 = g(t_1)$ and $A_2 = g(t_2)$ are the intersection points (we assume that g intersects the circle at two points), then $t_1 t_2 = b = |P|^2 - r^2$. From $|P - A_i| = |P - g(t_i)| = |t_i v| = |t_i|$ we obtain

$$(*) \quad |P - A_1||P - A_2| = |t_1 t_2| = ||P|^2 - r^2|.$$

This value is independent of v and g.

(b) Let A and B be the endpoints of the two long rods of the inverter, where A is the center of the dotted circle of radius r in the figure for Exercise 42(b). We consider the points 0 and A as fixed in the plane. Then x and Fx always lie on the dotted circle. Moreover, since 0, x and Fx are at equal distances from A and B, they lie on the perpendicular bisector of the line segment

[4] The same is true for any polynomial equation $t^n + a_1 t^{n-1} + \ldots + a_n = 0$: Denoting the solutions $t_1, \ldots, t_n$, we have $t^n + a_1 t^{n-1} + \ldots + a_n = (t - t_1)(t - t_2)\ldots(t - t_n) = t^n - (t_1 + \ldots + t_n)t^{n-1} + \ldots + (-1)^n t_1 t_2 \ldots t_n$ for *all* t; thus, by coefficient comparison, it follows that $a_1 = -\sum_i t_i$ and $a_2 = \sum_{i<j} t_i t_j$ up to $a_n = (-1)^n t_1 t_2 \ldots t_n$. These expressions in $t_1, \ldots, t_n$ are called the *elementary symmetric polynomials*. Thus, given the solutions $t_1, \ldots, t_n$ of such an equation, we can easily compute its coefficients $a_1, \ldots, a_n$. Algebra deals with the inverse task: finding the solutions from the coefficients.

$[A,B]$, so they are collinear. Now we can apply (a) with $P = 0$, $A_1 = x$, $A_2 = Fx$. However, in (a) the origin 0 was chosen as the center of the circle; for another center point A the formula $(*)$ is changed to $|P - A_1||P - A_2| = ||P - A|^2 - r^2|$. Using $P = 0$ and $|A| = s$ we obtain $|A_1||A_2| = s^2 - r^2$. Thus the map F being realized by this mechanical device is the inversion along the circle with center 0 and radius $R = \sqrt{s^2 - r^2}$.

43. Clearly, the composition of inversions along two spheres of radius r and s with common center 0 is a homothety: According to Eq. (6.5) we have $x \mapsto y := (r^2/|x|^2)x \mapsto (s^2/|y|^2)y = (s/r)^2 x$. The general case is reduced to this particular case using a suitable inversion, as shown by the following figures.

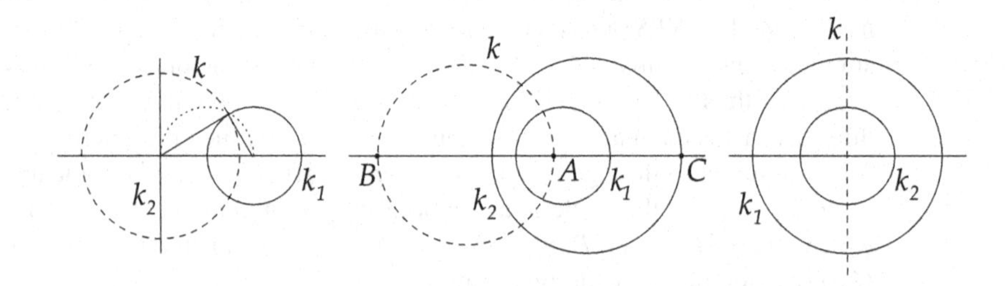

First we look at the middle figure and construct the dashed circle k intersecting both circles k_1 and k_2 perpendicularly. By an inversion along a circle centered at C (the right intersection point of k_2 with the horizontal axis) we transform the figure such that k_2 becomes a vertical straight line. After reflection along this line we obtain the left figure. Using the dotted Thales circle we easily find a circle k intersecting both the circle k_1 and the vertical line k_2 perpendicularly. By transforming back we obtain the circle k in the middle figure.

The inversion F along any circle with center B maps B to ∞ and thus k onto a straight line while preserving the horizontal axis. Thus k becomes a vertical line intersecting both circles k_1, k_2 perpendicularly (right figure).[5] Now this figure is mirror-symmetric with respect to both the horizontal and the vertical axis, hence the circles k_1 and k_2 have common center at the origin which is the intersection point of the two axes. Thus we are back in the special case discussed at the beginning.

This inversion F transforms the radial rays in the right figure below onto the circular arcs between A and B in the left figure. They intersect k_1 and k_2 perpendicularly and therefore they remain invariant under the inversions along both k_1 and k_2.

[5] Observe that the inclusion of k_1 and k_2 (k_1 inside k_2) is interchanged by inversion.

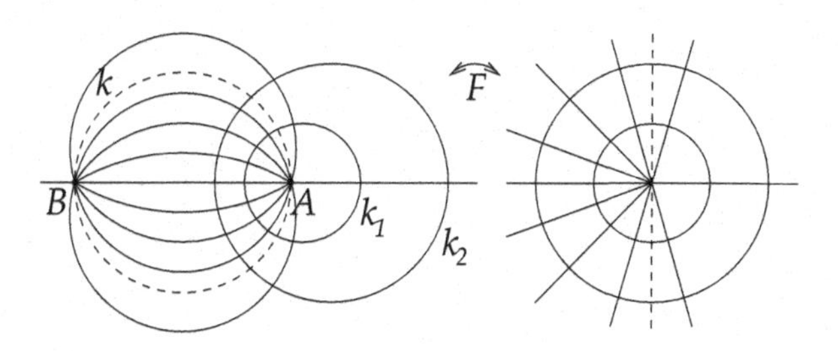

44. $4\pi = 2F + 2(\pi - \alpha + \pi - \beta + \pi - \gamma) \Rightarrow F = \alpha + \beta + \gamma - \pi$.
45. We have for the cross-ratios:

$$(v, \tilde{w}; n_1, n_2) = (0, t; -1, 1) = \frac{-1-0}{-1-t} : \frac{1-0}{1-t} = \frac{1-t}{1+t}.$$

Because $t = s/c$ with $s := \sinh a = \frac{1}{2}(e^a - e^{-a})$ and $c := \cosh a = \frac{1}{2}(e^a + e^{-a})$, we obtain

$$\frac{1-t}{1+t} = \frac{c-s}{c+s} = e^{-a}/e^a = e^{-2a},$$

thus $\frac{1}{2}|\log(v, \tilde{w}; n_1, n_2)| = a =$ hyperbolic distance.

Literature (Small Selection)

Mainly Used Literature

1. Marcel Berger: Geometry I, II. Springer Universitext 1987
2. D. Hilbert, S. Cohn-Vossen: Geometry and the Imagination. Reprint AMS 2020

Classics

3. H.S.M. Coxeter: Introduction to Geometry. 2nd edition. Wiley 1969, www.pdfcoffee.com/coxeter-introduction-to-geometry-pdf-free.html
4. B. Grünbaum, G.C. Shephard: Tilings and Patterns. Freeman 1987
5. R. Courant, H. Robbins: What is Mathematics? Oxford University Press 1996
6. Ebbinghaus et al: Numbers. Springer 1991

Historical Works

7. Euclid: Elements of Geometry. http://farside.ph.utexas.edu/Books/Euclid/Elements.pdf
8. Albrecht Dürer: Underweysung der Messung mit dem Zirckel und Richtscheyt, in Linien, Ebenen unnd gantzen corporen. Nürnberg 1525, https://proofwiki.org/wiki/Book:Albrecht_D%C3%BCrer/Underweysung_der_Messung_mit_dem_Zirckel_und_Richtscheyt
9. David Hilbert: Foundations of Geometry. https://math.berkeley.edu/~wodzicki/160/Hilbert.pdf
10. Bell, E.T.: Men of Mathematics. Fireside 1937/1965, https://archive.org/details/in.ernet.dli.2015.59359/
11. C.J. Scriba, P. Schreiber: 5000 Years of Geometry. Birkhäuser 2015
12. J.-H. Eschenburg: Sternstunden der Mathematik. Springer Spektrum 2017

Geometry and Art

13. D. Clévenot, G. Degeorge: Ornament and Decoration in Islamic Architecture. Thames & Hudson Ltd. (2000)
14. Martin Kemp: The Science of Art. Optical Themes in Western Art from Brunelleschi to Seurat. Yale University Press 1990
15. Emil Makovicky: Symmetry: Through the Eyes of Old Masters. De Gruyter 2016

J.-H. Eschenburg, *Geometry – Intuition and Concepts*,
https://doi.org/10.1007/978-3-658-38640-5

Some Textbooks

16. I. Agricola, T. Friedrich: Elementary Geometry. AMS 2008
17. R. Fenn: Geometry. Springer 2007
18. M. Koecher, A. Krieg: Ebene Geometrie. Springer 2007
19. M. do Carmo: Differential Geometry of Curves and Surfaces. Dover Publications 2016
20. A. Pressley: Elementary Differential Geometry. Springer 2010
21. J.-H. Eschenburg, J. Jost: Differentialgeometrie und Minimalflächen. Springer Spektrum 2014

Index

A
Action, 12, 30, 115
Additive, 13
Affine basis, 18
Affine coordinates, 19
Affine group, 117, 122
Affine independence, 19, 117
Affine mapping, 66
Alberti, L.B., 23
Algebra, 130
Alternating group, 73, 75
Altitude, 118, 141
Angle, 62
Antiholomorphic, 97
Antipodal, 28
Antipodal map, 74
Approximation, 9, 45, 60, 81, 85, 86, 89, 98, 119, 132, 149, 154
Arc length, 109
Area, 60, 78, 90, 118, 136, 137, 141
Art, 1, 23, 77, 120, 125
Axiom, 3, 4, 9, 35, 45, 60, 112

B
Bending, 90
Biangle, 137
Bolyai, J., 112
Boy Surface, 29
Braid relation, 93
Brianchon, C.J., 42, 47, 124
Brunelleschi, F., 23

C
Cardinality, 116, 140
Cartan, É., 5
Catenoid, 154
Cathetus, 58
Cauchy-Schwarz inequality, 63
Cell
  120-, 72
  24-, 72, 129
  600-, 72
Center, 30
Center of gravity, 19, 71, 118
Center of rotation, 76
Central perspective, 21, 120
Central projection, 23, 33
Characteristic, 11
Chord theorem, 102, 135
Circle, 48, 61, 123
Circumcenter, 118
Circumcircle, 119
Cocube, 71, 129, 150, 151
Collinear, 8
Collineation, 10, 29, 122
Complex differentiable, 96
Concepts, 4
Conformal, 96, 99
Conformal factor, 96, 134, 156
Conic section, 37, 48, 81, 124
Connectivity, 86
Construction, 58
Continuity, 29
Continuous, 16
Continuously differentiable, 87
Convex hull, 70
Convexity, 70
Coplanar, 44
Corner, 70
Cosine, 63
Cross product, 88
Cross-ratio, 53, 125, 137
Crystal, 77
Crystallographic restriction, 77
Cube, 70, 71, 73, 74, 79, 116, 128, 129, 139, 150, 151
Curvature, 2, 87, 89
Curve, 37, 87

J.-H. Eschenburg, *Geometry – Intuition and Concepts*,
https://doi.org/10.1007/978-3-658-38640-5

D
Dandelin, G.P., 81, 131
Definition, 2, 45, 62
Dense, 16
Derivative, 86, 96, 154
Desargues, G., 25, 33, 35, 46, 117
Diffeomorphism, 87, 96
Differentiable, 86
Differential, 86
Differential equation, 153
Differential geometry, 85
Dihedral group, 78
Dihedron, 76, 78
Dilatation, 10, 11
Direction, 10
Directrix, 83, 84
Discrete, 76
Distance, 4, 60, 61
  hyperbolic, 113, 137, 159
Dodecahedron, 70–73, 75, 79, 128
Duality, 45, 46, 72, 73
Dual space, 46
Dürer, A., 24, 120, 126

E
Edge angles, 71
Einstein, A., 90, 103
Elementary matrix, 51
Elementary transformation, 50
Ellipse, 79, 81, 83
Equation, 86
Equivalence relation, 25, 116
Eratosthenes, 86, 132
Erlangen program, 4
Euclid, 3, 4, 57, 58, 61, 64, 112
Euclidean group, 66, 76
Euler, L, 118
Euler line, 118

F
Face, 70
Fibonacci numbers, 119
Field, 9
Field automorphism, 14
Figure, 2
Flag, 70
Focal point, 80, 155
Formalization, 3
Fractional-linear, 103, 108, 114, 122, 148
Fundamental forms, 88

G
Gauss curvature, 90
Gauss map, 88
Gauss, C.F., 88, 90, 112
Generating line, 37
Geometry, 1, 10, 11, 58, 62, 65, 81, 90, 107, 112
  affine, 7
  conformal, 95
  exceptional, 65
  four-dimensional, 71
  hyperbolic, 108
  Lie, 105
  metric, 65
  non-Euclidean, 112
  polar, 65, 104, 105
  projective, 25, 35, 65
  spherical, 104
  symplectic, 65
Glide reflection, 68
Golden ratio, 70, 72, 75, 119, 120, 125, 128
Gradient, 133, 154
Grassmann manifold, 27
Gravity line, 118
Great circle, 29, 110
Group, 12, 30, 76
  classical, 36
  exceptional, 36

H
Half space, 70
Harmonic division, 55
Helmholtz, H.v, 65
Hexagon, 42
Hidden, 2
Holomorphic, 96
Homeomorphism, 16
Homogeneous polynomial, 38
Homogenization, 38, 105, 123
Homothety, 9, 10, 35, 102, 116, 118, 136
Hopf circle, 84
Horizon, 22, 48
Hyperbola, 80, 82, 84, 123
Hyperbolic line, 111
Hyperbolic plane, 109
Hyperbolic space, 109
Hyperboloid, 45, 108, 124
Hyperplane, 12, 45, 66
Hyperplane at infinity, 27, 28, 31, 100
Hyperplane reflection, 66
Hypersurface, 87
Hypotenuse, 58

I
Icosahedron, 70–75, 78, 119
Image, 86
Immersion, 87
Incidence, 4, 7, 21, 35
Inscribed angle theorem, 49, 149
Intuition, 1, 11
Invariant, 13
Inverse function theorem, 86, 92, 154
Inversion, 97, 98, 101, 102, 104, 135, 157, 158
Inverter, 135
Involution, 97
Islam, 77
Isometry, 65–68, 73, 76
Isotropy group, 116

J
Jacobian matrix, 86
Jacobi, C.G.J., 86

K
Kant, I., 2
Kernel, 46, 86
Killing, W., 5
Klein, F., 4, 112
Kronecker, L., 90

L
Length, 60, 109
Lie, S., 5, 105
Light cone, 103, 107, 137
Line at infinity, 25, 29, 37
Line segment, 8, 67
Linear, 14
Linear form, 46, 154
Liouville, J., 99
Lobachevski, N.I., 112
Lorentz group, 103, 107
Lorentz, H., 103
Lorentzian reflection, 104
Lorentzian scalar product, 103, 109
Lorentzian submanifold, 105

M
Masaccio, 23
Matheon, 119
Mean curvature, 90
Means, 19
Median, 118
Mercator, G., 133
Metric, 32
Minimal surface, 90, 133
Möbius, A.F., 104
Möbius geometry, 104
Möbius group, 104
Möbius strip, 29, 144
Möbius transformation, 103, 114, 148
Moon, 115
Motion, 66

N
Norm, 60, 61, 65
Normal form, 38
Normal space, 87
Normal vector, 72, 87, 88, 127, 150, 155
North pole, 100

O
Obvious, 2
Octahedron, 70, 71, 73, 75, 79, 119, 129, 142, 151
Octonions, 35, 65
Open subset, 86
Operation, 12
Orbit, 13, 27, 116
Order, 76
Orientation, 66, 96, 144
Oriented, 29, 66
Origin, 7
Orthocenter, 118
Orthocircle, 114
Orthogonal, 62, 91
Orthogonal group, 65, 66, 76, 77
Orthogonal hypersurface system, 92
Orthogonal map, 65, 66
Orthonormal basis, 64, 65

P
Pappus, 70, 116, 124
Parabola, 83, 123
Paraboloid, 83
Parallel class, 25, 26, 146
Parallel displacement, 10
Parallel map, 13, 16
Parallel projection, 16, 18, 23
Parallelism, 7, 11, 112, 119
Parallelogram, 7, 13
Parameter change, 87, 89
Parametrization, 86, 87
Pascal, B., 47, 48
Pentagon, 70, 75, 125, 126

Permutation matrix, 51
Perpendicular, 62
Perpendicular bisector, 67, 71, 118, 129
Phidias, 125
Photo, 25, 120
Plane, 12, 64, 76
Plane at infinity, 25
Plane of symmetry, 73
Plato, 70
Platonic group, 76, 79
Platonic solid, 70, 72
Poincaré, H., 113
Point, 12, 19
Point at infinity, 25, 105, 123, 146
Polarity, 48, 104, 105, 124
Polarization, 39, 65, 147
Pole, 103, 108, 124
Polytope, 70
Poncelet, J.-V., 25, 45
Preimage, 86
Principal axis, 79
Principal curvature, 89
Projection, 18
Projection center, 23
Projection line, 18, 23–25
Projective closure, 38, 39, 105
Projective group, 30, 108
Projective line, 122
Projective mapping, 30, 121, 123
Proof, 2
Proportional, 27
Pythagoras, 3, 20, 57, 70

Q

Quadratic completion, 50
Quadratic form, 39
Quadric, 38, 39, 44, 48, 49, 92, 103, 104, 123, 133, 146, 154
Quadrilateral, 54, 121
  complete, 54
Quantity, 20
Quasicrystals, 77
Quaternions, 11, 16, 35, 130

R

Ratio, 20, 52
Ray theorem, 9, 141
Regular, 71
Relativity, 95, 103, 107, 108, 114
Riemann, B., 90
Riemannian geometry, 90
Riemannian metric, 90
Rotation, 60, 73, 76
Rotation group, 73–75

S

Scalar product, 61, 64
Screw motion, 68
Segment, 64
Segre, B., 42
Self-dual, 72
Semi-affine, 15, 19, 20
Semidirect product, 66, 122, 141
Semilinear, 15, 122
Semiprojective, 30
Similarity, 59, 101, 125, 130
Simplex, 71–73, 117
Simply transitive, 12, 74
Skew, 35
Space
  affine, 12, 27
  projective, 26
Spacelike, 103
Special unitary group, 130
Sphere, 28, 77, 97, 100, 103, 104, 107, 108, 131, 133, 136, 138, 156–158
Spherical, 97, 99
Spherical coordinates, 91, 92, 99, 110, 111
Spivak, M., 88
Stabilizer, 116
Star, 70–72
Stereographic projection, 84, 100, 134, 135
Straight line, 12, 27, 124
Straight map, 29, 54
Subalgebra, 130
Subspace
  affine, 11
  projective, 27
Substitution, 50
Surface, 87
Symmetric group, 73
Symmetry, 3, 4, 7, 44, 73, 92, 129, 148, 151
Symmetry group, 70, 74

T

Tangent, 45, 47, 81, 101, 123
Tangent plane, 43, 45, 124
Tangent space, 45, 87, 124
Tetrahedron, 70, 71, 73, 74, 79, 117, 127, 150
Thales, 49, 98, 158
Theaetetus, 70
Timelike, 109
Tits, J., 5, 65
Torus, 84

Transitive, 12
Translation, 10, 12
Transpose, 61
Transposed, 130
Transversal, 27, 145
Triangle, 117, 118
Triangle inequality, 63

U
Umbilic hypersurface, 90
Umbilic point, 90, 99
Unit, 20
Unitary group, 130

V
Vanishing line, 22, 143
Vanishing point, 21, 143
Vector, 8, 19
  homogeneous, 27, 45
Vector group, 12
Vector space, 8, 9
  Euclidean, 64
Vertex, 70

W
Weingarten map, 89
Weingarten, J., 89
Weyl, H., 5, 65